Small Cattle for Small Farms

SECOND EDITION

Margo Hayes

CSIRO PUBLISHING

National Library of Australia Cataloguing-in-Publication entry

Hayes, Margo, author.

Small cattle for small farms / Margo Hayes.

Second edition.

9781486301867 (paperback)
9781486301874 (epdf)
9781486301881 (epub)

Includes bibliographical references and index.

Cattle – Australia.
Beef cattle – Australia.
Farms, Small – Australia.

636.2130994

Published by
CSIRO Publishing
36 Gardiner Road, Clayton VIC 3168
Private Bag 10, Clayton South VIC 3169
Australia

Telephone: [+613] 9545 8555
Email: csiropublishing@csiro.au
Website: www.publishing.csiro.au

Front cover (clockwise, from top left): Ausline bull (courtesy of Southern Star Elite Ausline), Vitulus cows in the paddock, red Ausline calf, Belted Galloway (courtesy of Australian Belted Galloway Association)
Back cover: The author with a Lowline bull

Set in 11/13.5 Adobe Minion Pro and Helvetica Neue LT Std
Edited by Adrienne de Kretser, Righting Writing
Cover design by Alicia Freile, Tango Media
Typeset by Desktop Concepts Pty Ltd, Melbourne
Printed by Ingram Lightning Source

Jun26_RP_ILS

Contents

Acknowledgements

I would like to gratefully thank and acknowledge a number of people who made this book possible. First my father, who sadly never got to see the second edition: it was he who gave me my love of livestock and who spent so much time helping the Vitulus Stud over the past 20 years. To my daughter Mercedes, who made it all happen, and my husband, whose support and encouragement were essential for its completion.

In addition, I would like to thank the relevant breed societies featured, which provided the information and many of the photos, and Dr Sandi Jephcott for her editing and assistance with Chapter 11, 'Nutrition'.

Introduction

Having been involved with small breeds of cattle, including Ausline and Australian Lowline, for the past 20 years, this book started as an information sheet to answer the many and various questions that I receive either via email or directly from my clients. The majority had no prior knowledge or experience with cattle and often what seemed obvious or second nature to me was, on reflection, a genuinely sensible question from their perspective.

This book is not here to extol the virtues of any particular breed, as all have their strong points. It's just a matter of finding which one suits your circumstances, climate, acreage size and objectives. For many it's not about making a profit, it's about fulfilling a lifetime wish of owning land and running some cows to manage the grass and provide a rural environment for children and grandchildren.

Small or miniature breeds fit these criteria perfectly for a number of reasons. Their size, ranging from around 100–130 cm, means they are not as daunting as some of the larger breeds that can come in at over 180 cm. When converting this to kilograms some of the smaller breeds will range from 300 kg to 700 kg at maturity compared to the larger breeds ranging from 700 kg to 1500 kg! Their stocking rate means their smaller size allows you to run more per hectare and produce more beef per hectare, and their docility makes them ideal first-time cattle.

My area of expertise is, as stated, in breeding stud Ausline and Australian Lowline cattle alongside a commercial herd, but the principles are the same for all small breeds. Each breed will have its own Association and requirements and you will need to contact the relevant breed society to find these out. I also suggest that you obtain a copy of the society's rules and regulations, as these may impact on your long-term goals.

1

Time for a tree change

My reason for making the tree change came about by accident. My husband decided, when our youngest was just a baby, to retire and write books so he bought a farm in the hinterlands of northern New South Wales, obtained a contract with a major publishing house and started writing. There was I, an events organiser of international jazz artists, used to being extremely busy, with a tiny baby stuck in the hills on a farm with not much to do. I explored all options for our 50 acres and I eventually chose small cattle. I had been brought up on a property of some 7000 acres, my father being a sixth-generation of Australian graziers, and I knew cattle. They seemed the obvious choice for me.

Then when our youngest was three she developed Type 1 diabetes, necessitating my care and attention 24/7, meaning that I could not return to the workforce with commitments to set hours for quite some time. So I decided to develop my cattle into something more than just a hobby. It has been the most wonderful time – I have been able to take and involve my daughter every step of the way. It's a lifestyle that keeps us outdoors and develops responsibilities in children in caring for animals, and has allowed us to travel together with a common interest. It has involved the whole family and now includes our grandchildren. In addition, it has allowed me to build a business through which I now export to seven countries, have my own award-winning branded beef product, and run introductory cattle care courses as well as producing stud and commercial animals. The opportunities are endless and the ability to create new markets by value adding to my core product is so rewarding. I have even seen the development and creation of a whole new breed which started out by isolation of a particular gene. It is challenging, exciting and what's more I absolutely love it.

So you have made the decision to buy some land, for whatever reason, and now you need some stock to eat the grass. Below are some questions you might like to consider.

Q. We have decided to do the tree change thing and buy some land. We would like to purchase between 50 and 100 acres. We like the idea of running a small breed of cattle as neither of us have had any previous experience with livestock. Do you have any suggestions for us prior to the purchase of our land, or factors that we need to consider in order to run our cattle?

A. The most important factor when looking at purchasing land is availability of quality water. You need to have access to a clean and constant water supply whether it be from creeks, bores or dams. If dam water is the only supply, then check to make sure the dams have never been dry. You can always buy in feed if need be, improve the pastures, build shade shelters and put in new fences but it can be almost impossible – and exorbitantly costly – to buy in water for livestock.

Q. I have recently bought a small farm for the sake of starting a wholesale nursery (I have been in the nursery industry for 14 years) but I will not use all my acreage for some time, if ever. I would like to find a product that could keep down the grass (as seed flying around is detrimental to my potted crops) and perhaps provide an additional income. What can you suggest? What costs are associated? What sorts of profit (if any) could I expect?

A. In order to gauge your profit margin you will need to determine whether you will run stud or commercial cattle. Other factors that need to be considered include whether supplement feed will be required and, if so, whether you can self-produce or will have to buy it in, what is the average rainfall, what is the condition of the soil and pastures and what is the livestock carrying capacity of the land. All these issues will have an impact on the bottom line. Bear in mind that it is highly unlikely the majority of small farms will ever make a substantial profit due to economies of scale, but it is possible to run enough cattle so that they cover their costs and, if the season and markets are good, provide you with a small profit.

Q. We have recently purchased a small farm (10 acres) and we are looking at getting some Dexters from a lady in our street who breeds them. I need knowledge, knowledge, knowledge and I was wondering if you could

point me in the right direction with how to go about learning about cattle, this breed and running them etc.

A. My research has shown that the majority of people who purchase smaller cattle have no prior knowledge. Obtaining information from breed societies and members and going to shows and field days are excellent ways to learn. Specific courses aimed at introductory cattle husbandry are also an excellent way to learn the basics. The various state government primary industry websites are valuable reference tools, along with the Meat & Livestock Australia (MLA) and FutureBeef websites.

Q. I have a house on 5 acres of cleared land (which was once cattle pasture), which I will be moving to full-time around October this year. I intend to try the 'self-sufficiency' thing (I've already done the basics, vegie garden, chooks etc.), and am looking into some form of livestock. My neighbour runs three beef cattle (also pigs, goats, sheep, ducks, chooks etc.) and I am interested in doing the same thing. I thought that if I had the 'smaller' cattle I could perhaps run a few more. Could you please give me some information on how many I could keep on 5 acres (with dam), approx. cost (not too worried about buying the next state champion!) and availability.

A. The recommended stocking rate for small cattle is one animal per acre, but this depends on the average rainfall, condition of the soil, pastures and tree density. You also need to consider how much available grazing land there is, taking into consideration the house yard, sheds etc. Cattle are herd animals so they like to have a companion. The MLA website has a stocking rate calculator which you might find useful but remember it is based on larger breeds.

Q. What type of grass is best for grazing (real beginner question here!)? I'm looking at optimising the block to cater for the cattle, and hope that my 5 acres will accommodate four to five cows and a bull.

A. Evidence would seem to suggest that cattle do best on a variety of grasses and pastures. You will need to find out from your local Department of Primary Industries (DPI) what grasses grow best in your region. It would be highly unlikely that 5 acres would be able to accommodate that number of stock if the cows are to be bred, which means that in nine months there will be nine to 11 head.

Q. I have just purchased a small (110 acre) property in NSW. I am a practical fellow, but with no agricultural experience. Ultimately I would want to go into beef farming rather than breeding.

A. Common practice is to buy in stock when pasture is plentiful and destock in times of dry or reduced pasture growth. This can be quite a profitable exercise if the seasons are good and the cattle markets are buoyant, but a word of caution – everyone else in the beef industry will be doing the same thing. My advice would be to start with a small number and then, once you are able to estimate the land's stocking rate during the various seasons, you can build your numbers.

Q. I suppose my plan is to have three or four cows and a bull and breed them in order to feed my family. I don't plan on getting so involved that I'm dealing with saleyards. I just want something to keep the grass down, and every so often get the local 'roaming butcher' in to slaughter one. Hopefully this is feasible.

A. This is a perfectly feasible plan. Small cattle are perfect for the freezer as they produce manageable cuts and a whole carcass is able to fit into a freezer. Remember that once killed a beast will weigh about 55% of its original live weight. From this weight you should then obtain 60–70% of useable meat once bone, fat etc. have been discarded.

2

Benefits of smaller cattle

There are many advantages of smaller cattle. Small cattle are easier to handle, their size makes them not so daunting for first-time livestock owners, and the equipment required is less expensive and complicated. Their size also means that they just don't have the weight and strength to do as much damage to gates, yards, equipment and fences, so less maintenance is required.

Smaller cattle are also gentler on the land. You have far less damage to pastures due to the hoof size and weight of the smaller animals. In countries such as New Zealand, where pugging of the soil is a problem in areas of excessive rainfall, small breeds have been found to be excellent in preventing this while at the same time allowing the pastures to be grazed effectively.

As the imperative to increase farm profitability continues, production costs must be minimised and the returns to the producer must be maximised. Smaller cattle offer a number of significant and obvious advantages such as higher stocking rates, ability to maximise available feed, shorter production time, lower production costs and higher-valued products. You can produce more beef per hectare. It takes about 2 hectares to raise two large breed animals whereas you can raise four per hectare with smaller breeds. On highly fertile land with a high rainfall, this stocking rate can be increased. There are more small farms in Australia than there are big farms, and small cattle are well suited to these. More importantly for the farmer, being able to produce finished cattle sooner means more cattle per hectare.

Small cattle are also well suited to the beef market: the Australian appetite is for a thick juicy portion of steak not for something that is 30 mm thick and hangs off the side of the plate. When it comes to beef production, small cattle have finer-grained beef; this is the most tender beef. Most of the smaller breeds have excellent meat qualities. The numbers that compete successfully in our national hoof and hook competitions support this, as do their significant wins in branded beef competitions.

The International Miniature Cattle Breeders Society and Registry claims that the small breeds are 25% more efficient in terms of feed conversion, requiring about a third less feed than their larger counterparts. One of the largest Angus studs in the US, the Duff Cattle Ranch, claims they are 30% more feed-efficient. As the imperative to produce grass-fed beef continues to grow, whether it be organic, biodynamic or just certified pasture-fed, the early maturity pattern of the smaller breeds gives them a commercial advantage. This means that they can stop growing at a younger age and express fattening ability at an earlier age so they can produce a finished product sooner.

Those that are marketed as a beef breed have shown excellent retail yield percentages with some butchers obtaining up to 75%, which is very impressive. Retail yield is the percentage of useable meat from a carcass. This is due to the fact that the smaller breeds don't have as much bone, especially leg bone, which is all wastage. This allows the butcher to make more profit per carcass.

Smaller birth weights are another advantage. This reduces the number of calving problems and it is why we now see a number of commercial producers of larger British breeds using smaller breed bulls over their heifers, hence the term 'heifer bulls'. Don't worry, I can assure you, they don't seem to have a problem reaching! I have had Charolais and Senepol cows in excess of 700 kg that I join annually to my Ausline bulls. Lower birth weight calves prevent the heifers from suffering damage during their first calving, allow them to recover and cycle again faster and conceive again sooner.

In addition, the ability of the smaller breeds to moderate the frame size of larger traditional breeds without compromising beef qualities optimises efficiency and sustainability. In more recent years producers have made a concerted effort to reduce frame size, add muscle and produce an animal that can fatten off grass. Exceptionally large cows have been shown to wean a lower percentage of their bodyweight with an unsustainable feed consumption requirement, as well as reducing fertility and longevity.

An increasing number of dairies, especially in the southern states, consider that small breed bulls provide a commercially attractive alternative to traditional choices for dairy heifers, for a number of reasons. First, due to ease of calving, heifers joined to smaller breeds cycle again quickly and hit full milk production rapidly after calving. This expedites their return to the dairy and significantly enhances the probability of taking AI (artificial insemination) on the first cycle. Second, they produce excellent quality beefy calves, which do extremely well in the traditional dairy calf markets. Third, management and safety issues are greatly reduced. Smaller breeds have temperaments that are preferable to those of traditional dairy bulls, which are renowned for being extremely dangerous. This allows dairy farmers to walk freely among their cows and bulls and manage them much more easily.

Finally, the most important benefit is they make great family pets. Children love them: they can pat the calves and milk the mum and generally interact with them with the greatest of ease. The cattle end up becoming part of the family, with names and personalities of their own. Most owners find the whole idea of using them for beef is abhorrent but remember that you can't keep everything you breed, especially bulls.

Q. Is 5 acres big enough for a Lowline bull and heifer and hopefully a calf in future?

A. Without knowing if there is irrigation available and whether the pastures are improved, let alone seeing a soil analysis, average rainfall, density of vegetation and temperature graph, it's very hard to say. But providing there is adequate rainfall with good pastures then in most circumstances it would be possible. You may have to supplement feed during the drier months or when grass growth is reduced or severely frosted.

My property allows me to run 130 head on 160 acres (including my commercial herd) but I have irrigation and improved pastures and I do have to feed hay and molasses supplement in dry times. We have a very dry winter with normally no rain from May to October so I need to plan in advance for those dry months; for example, I grow my own hay.

3

Production systems: stud or commercial

The first question I ask any potential buyer is 'What is your purpose in buying cattle?' This will determine whether you want stud or commercial stock. Are you seeking to make a profit or will it just be a hobby? Do you want to show or do you just want to produce grass-eaters to fatten and sell through the markets, or perhaps even home kill or sell at farmers' markets? Do you want to breed your own animals? By far the easiest types of stock to manage are steers but then you don't have the enjoyment of breeding and producing baby calves. If you go down the track of commercial animals, where do you intend to sell the progeny? Management issues also need to be considered; for example, if you are unable to check your stock regularly then fattening steers maybe a more suitable option. You need to be able to answer these questions before you decide upon purchase of your initial stock because it can be a considerable capital investment and you will want to spend your dollars wisely. Once you begin, whichever way you go, it's going to be a very steep and rapid learning curve.

I suggest that you start off in a small way. Don't go in and buy a whole herd! Instead, start off with a few and learn gradually. Then when you feel more confident you can expand.

Q. What is the difference between stud stock and commercial stock?

A. **Stud stock** are purebred or fullblood stock and are registered with the relevant breed society. They normally command a higher price. It will therefore take a lot longer to obtain a return on your investment, as the majority of females will be kept until your breeding numbers are built up.

Some breed associations allow grading up to purebred status. This means that each successive generation has a higher percentage of that particular breed until you get to purebred status, which is generally achieved in four or five generations. You will need to check with the relevant breed society and its rules.

Stud stock-owners should guarantee their animals for sale and provide you with detailed health and joining records as well as registration papers. Many breed societies also request that all registered animals are DNA typed for parent verification. Registered stud stock are the only stock in Australia that are able to be exhibited at shows, with the exception of led steer classes. Animals that exhibit traits and characteristics that are considered desirable by judges may prove more expensive to purchase. There are a lot more administration costs associated with stud animals due to registration, transfer and membership fees.

Commercial stock are either crossbred animals or unregistered purebreds. Crossbred cattle are a mix of breeds with varying percentages (i.e. they are not purebred or registered) and as such are usually cheaper to purchase. Commercial stock cannot be exhibited at shows and, if they are of mixed breeds, can often grow to be larger animals than purebred stock. This will depend on what breed they have been crossed with – obviously a Charolais/Ausline cross will be a bigger animal than a Dexter/Ausline cross. If you just want small quiet grass-eaters then commercial stock are often a sensible and cheaper option. Some breeders also sell unregistered purebred animals as commercial cattle. They are normally unregistered because the breeder does not feel they are of significant quality to be registered.

Q. We want to have a small breed but we are still unsure whether to have stud or commercial cattle. We do want to breed. Do you have suggestions?

A. Depending on the acreage available, my advice in this situation is to start off with some commercial heifers (1st cross), a registered bull and one or two stud animals. The bull can be joined to both the stud and commercial females. The commercials will give you a more immediate cash flow as their progeny can be sold, while you will probably retain your stud animals until your breeding numbers build up. This will give you some time to see which avenue you wish to pursue without outlaying a large sum of capital in stud stock.

Q. I think Lowlines are the cattle for my property. I am wanting to sell them for commercial reasons only, I am not interested in showing. I intend to breed with only well-bred stock. My question is, do you think Lowlines are sellable to the general public? Will they sell as well as the bigger breeds? My husband doesn't think they are the type to just take to a cattle sale. What are your opinions on this? Is it easy to sell to butchers as a meat they would want?

I have been looking at getting Lowlines for a while now but after talking to other cattle farmers they say that butchers will not take the Lowline or Dexter breeds for meat. They say to stick to Murray Greys and Angus. Why is this, and is there any truth to the story or is it just schoolyard talk?

A. I've heard these comments a hundred times. In recent years the quality of meat from the smaller breeds has gained national recognition in both carcass competitions and branded beef awards. It's not that butchers won't take on the smaller breeds, it's the saleyards and the agents that don't like them and don't support them. If your only market to sell your stock will be the saleyard then purebred or fullblood smaller breeds will not give you the return you deserve. If you can find your own direct market, such as selling directly to the butcher, then you will financially be a lot better off. Having said that, I have had a number of clients top the local saleyards with their 1st cross progeny, based on size of the animals presented to the market.

Q. I was advised today by the stock and station agent we are purchasing the property through (who admitted to knowing next to nothing of this breed) that he would put six to eight Angus or Herefords on the property, simply due to ease of turnover and $ at the end of the day. His main advice to me was, whatever way I went, to look at the exit strategy(s).

A. This is typical agent speak and he probably runs the local saleyards so wants another sale. There is no attempt to understand the smaller breeds or the fact that his client has no prior knowledge of cattle. Six to eight unruly cattle from the saleyard would be a total disaster. As discussed in this book, there are numerous other exit strategies rather than the saleyard.

Q. Even though I have only a small hobby farm – 10 acres – I'm interested in breeding (very small scale). Would it be worth considering or should I just go with some cattle as pets? If breeding is an option, I would need advice on selecting a suitable mating etc. So, I guess what I'm trying

to say is that I don't know whether to go with the purebred or the commercial cattle.

A. This person was a typical client who had no prior knowledge of cattle. They bought a stud heifer and a commercial cow and calf. Starting with just a few animals allows you to gauge your carrying capacity as well.

Q. By way of background, my wife and I have recently purchased 16 acres near Ipswich. We both grew up on properties and have purchased the property to give our five- and three-year-old boys a taste of a country lifestyle even while work ties us close to the city. We are keen to ensure that we can produce as much food as possible for ourselves, and have been investigating cattle breeds that might suit our needs. We have been given your beef and were impressed. After perusing your website I saw the Senepol crosses and have done some reading and are leaning towards them at this stage. So now that I've prattled on, would you suggest that they would be suitable for us and if so, based on what stocking rates?

A. As this person was interested in purchasing livestock only for the family's meat production and the interest is primarily meat quality, then it would be perfectly reasonable to purchase 1st cross or commercial animals for this venture.

Q. What is the best way to start? e.g. a cow in calf, or a cow with a calf? Or is there another way?

A. There are a number of ways to start your involvement with a breed, and your production method may be constrained by monetary considerations as well. Here are some options.

- Buying pregnant females is the fastest way to start. Make sure they are guaranteed in calf; if you are buying from a reputable breeder this will not be a problem. Sometimes you may buy a three in one package, i.e. a cow with calf at foot and in calf again.
- Buying young unjoined females and a bull. This will give you nine months before you have progeny on the ground and allow you time to get used to your cattle and handling them before you have to deal with calves.
- Artificial insemination (AI). If you don't want to own a bull then you may consider using AI on your females. You will need someone who is an experienced AI technician and does it regularly – they have the best conception rates. Dairy farmers make excellent technicians as many do

it on a daily basis. You or your technician will need an AI tank within which to store the semen, and preferably someone close by to attend both morning and afternoon when the females are cycling. You will also need to have the right facilities, such as a crush with vet access to restrain the animal.

- Embryo transplants (ET). I would not suggest this process to anyone who is starting out with cattle for the first time. There are too many factors to take into consideration, which require detailed cattle experience. It can also be a very expensive process with sometimes very poor results even for those who do it on a regular basis, I can assure you.

Q. I don't really want to keep a bull. Our place is too small. Is it hard to AI cows?

A. Learning to AI is not a difficult skill and there are a number of courses that private technicians, TAFE and agricultural colleges run. Attention to detail is the key to AI success, as it is all about timing and making sure your cattle are at the correct point in their cycle to AI. You will need to purchase the equipment, a storage tank for the semen and liquid nitrogen. An alternative option is to lease a bull for a number of months from a breeder.

Q. How much would we expect to pay per animal, where do we buy them and/or should we just have heifers and have them inseminated?

A. Price I can't comment on because every breeder and breed will charge accordingly. The breed society will advise you of breeders in your area. Getting animals inseminated (also referred to as AI) is an option for many who don't want to have a bull, but it can be costly both financially and in time. You will need experience to know when your females are cycling and have the time to observe and record when this occurs. As mentioned previously, you will need to have the semen stored close by so that you have access to it when it is needed.

Q. I'm interested in buying a single animal, as a gift, basically as a grass-eater and companion for one of my mother's old horses. I have done a fair bit of research on small breeds and they seem to be ideal for our purpose here. I would probably be looking to buy a steer, depending on how early you castrate them. But basically, I just need a reasonably cheap option, and we are not buying to breed or eat, so top bloodlines are not essential. Although I would prefer a male, an idea of both female and male prices

would be much appreciated. I'm pretty much just looking for a nice quiet animal, that isn't likely to turn nasty, because my mother has had little experience with cattle.

A. Cattle are herd animals and as such prefer to be with other cattle. The perfect choice for this person would be steers as they are the easiest to manage. I know of some people who ended up teaching their three steers to be ridden and they became very much a part of the family.

Q. I'm interested in the breeding side of things. I realise that due to the small size of my home that I wouldn't be able to have large numbers of cattle. Is breeding even on a small scale, even an option?

A. Breeding is very definitely an option on a small scale and a very enjoyable one at that. Nature is very good at doing its job so all you need is a female and a bull.

Q. We are considering buying some embryos to start our herd. As we are novices when it comes to breeding cattle, what is your advice and how do we go about it?

A. My advice is don't do it until you are much more familiar with the breeding cycle of cattle. You need to be able to know how to detect a cycle and the strength of a cycle because a successful ET program is all about getting the timing absolutely spot-on. When I first became involved, I made the mistake of trying to run my own ET program. I thought it would be a good way to fast-track the process and quickly build up my herd. In reality, I lost 12 months of time and would have been far better off putting the money into buying pregnant females. The heifers that I bought to use as recipient cows were too young. I was advised by one vet to buy maiden heifers but the ET vet wanted to use cows that had already had a calf, as maiden heifers are more difficult to implant. Having gone to the expense of buying good-quality Jersey heifers because Jersey cows have the highest butterfat content, which would be beneficial for the embryo calves, I then had to join them to a commercial bull and wait 12 months until they had calved. Out of the 12 only 10 could be implanted and only six embryos stuck. I can't remember what progeny I ended up with, but the time and money spent would have been far better invested in buying live cattle. One breeder I heard of bought seven pregnant recipient cows, when he first started out, and ended up with seven bull calves! Technology has advanced now and you can buy sexed embryos but it can still be an expensive process.

4

Selecting a breed

There are over 250 breeds of cattle and over 50 recognised breeds of small or miniature cattle throughout the world. Classification as a small breed means that the breed is governed by a height regulation; every breed has different interpretations of how these restrictions are applied.

Once you have decided on a breed, gather as much as information as possible. Contact the relevant breed societies and get a list of breeders in your area, visit a number of different studs and compare, go to local shows and view the cattle being judged. Talk to the breeders and to people new to the breed, ask who they bought from and why. If you are buying stud stock, the integrity of the breeder is paramount.

Role of the breed society

The role of the breed society is to preserve the genetic heritage of the breed and provide support for its members. Animals are registered in the respective herd book and their pedigree recorded accurately. All registered stud stock animals are issued with a certificate detailing the animal's pedigree, usually to five generations. Registered stud stock allows members to ensure accuracy of genetics, especially now that a number of breed societies enforce DNA-testing for parent verification. Societies may provide further support for members in a number of different ways, including marketing support such as brochures and flyers, information kits, website advertising, breed shows, promotion at shows and sales and a whole host of other services.

Breed types

The breeds of cattle found throughout the world originate from two main gene groups: *Bos indicus* and *Bos taurus.*

Bos indicus cattle, also referred to as tropical cattle, are derived from Asia and Africa and are distinguished by their hump, loose skin, fine hair and floppy ears. They are more suited to tropical conditions due to their insect, heat and parasite resistance and are found predominantly in the northern areas of Australia. Breeds include Brahmans, Santa Gertrudis, Droughtmasters, Bramalows and Nadudana.

Bos taurus cattle originated from Europe and are known for their meat and carcass qualities. They are divided into two divisions – British and European.

British cattle are generally more early-maturing and are therefore able to fatten sooner. They include Herefords, Miniature Herefords, Angus, Ausline, Australian Lowline, Dexters, Murray Greys, Galloway and Square Meaters.

European breeds are generally faster-growing, have more muscle and are later maturing. They include Charolais, Limousin, Bazadais and Simmental.

The following information on the small breeds included in this book has been provided and written by the relevant breed society. They are listed alphabetically.

Ausline

The Ausline Cattle Association Inc. began in 2002 under the name of Redline. In 2012 a name change took place to expand its objectives to become a viable commercially oriented alternative small breed, enabling the Association to meet the diversifying demands of smaller acreage producers. The Association has members in all the eastern states of Australia as well as a number of other countries.

Ausline calf.

Ausline bull.

The breed's ability to produce pasture-reared beef with a shorter growing time, due to its early maturity patterns, are perfect for those seeking to grow a environmentally sustainable breed, allowing optimum returns per hectare. A number of members are organic, biodynamic and pasture-fed accredited. Ausline Beef has won a number of national awards, not only in branded beef taste competitions but also Grand Championships in carcass competitions, which is testament to the beef quality.

The breed's beef qualities combined with its size, docility, quiet temperament and small birth weight makes them the ideal choice as nice quiet grass-eaters for a hobby, stud stock for those wishing to show and breed, or commercial animals to make a financial return.

The Association is distinctively different in that:

- it has a percentage and a fullblood register;
- upgrading to purebred status is allowed;
- it includes the Native Angus in its herd book;
- purebred and fullbloods can both be shown together;
- red cattle and black cattle can be shown;
- all coloured cattle can be accepted into the register;
- there is a height limit of 130 cm for the show ring only;
- all animals in both registers are DNA-typed for parent verification;
- fees and administration costs are very low in comparison to other breeds;
- there is a moderator program for commercial viability.

A fullblood herd is maintained with DNA parent verification that is required to ensure the purity of the herd. All fullblood animals can trace their parentage to the original Angus herds imported to Australia.

Australian Lowline

Australian Lowline beef cattle were bred and developed from the famous stud Angus herd that was established by the NSW Department of Agriculture at the Trangie Research Centre in 1929. In 1963 the emphasis at Trangie moved from breeding to scientific research. In 1974, trials to evaluate selection for growth rate on herd profitability commenced. After 15 years of selective breeding, one outcome was a smaller Low Line herd. In 1992 some of these cattle were purchased from Trangie, and the Australian Lowline Cattle Association was formed.

The range of markets for the Lowline beef breed enhances economic viability for the small acreage farmer. This breed's quiet temperament and small size, and the facts that it's naturally polled and has no inherent dwarfism gene, make it an affordable alternative to other breeds of cattle. Low birth weights ranging around 25 kg provides easy calving and a higher percentage rate of cows returning in calf. At all stages of their growth Lowlines are smaller than Angus. The dams make excellent fertile mothers and provide ample milk for fast-growing calves.

Lowlines can be run at a commercially viable stocking rate nearly twice that of most other beef breeds, and be finished on grass. Lowline bulls can also be crossed successfully with other breeds, resulting in quality moderate-framed commercial cattle (see photo in colour section). Cattle produce tasty, tender and evenly marbled beef with carcasses producing exceptionally high retail yields.

Lowline cattle thrive in a wide range of environments from the snow areas to the tropics.

Lowlines are a closed herd book breed, providing genetic traceability to the original Angus herd at the Trangie Research Centre. Upgrading to purebred status is not permitted.

The Lowline Association embraces about 300 registered breeders in Australia, New Zealand, the US, UK and Canada. It encourages breeding for a diverse range of viable markets, including quality beef production, stud development, and bulls to support dairy and other beef breed operations. Certified Lowline Beef is developing as a member-driven, traceability assurance arrangement to provide consumer confidence in Lowline beef. Member activities include showing stud Lowlines at royal, regional and local shows, beef carcass competitions, Lowline promotion at industry events and interaction of Association members through promotion groups. Expanding export opportunities will build on existing Lowline cattle and genetic exports to Europe, the UK, Asia, North and South America.

Boran

Borans are a very ancient cattle breed. They were highly prized by the Maasai tribal people of Africa, who bled them from the neck and combined the blood with milk for consumption. Due to their incredible hardiness and versatility, Kenyan ranchers recognised Borans as the best indigenous beef animal in East Africa and so it became the dominant breed in the region.

Boran bull.

A team from CSIRO in Rockhampton, led by Dr John Frisch, recognised the value of introducing a tropically adapted breed into Australia, that was unrelated to the Brahman. Their research identified the Borans as having the characteristics required.

Borans are a medium to small beef breed. Colour combinations can vary between white, grey/black, fawn and red. They are recognised for their high fertility, good mothering ability, excellent temperament and great survivability under harsh conditions. They possess a very high resistance to ticks and internal parasites. They calve very easily and show a much more gentle nature than other *Bos indicus* cattle. Their wonderful temperament and meat quality separate them from other tropical breeds. Borans are both horned and polled (no horns).They are not all quiet instantly, but with their inquisitive nature and a little attention you will be amazed at their response. Their tolerance of harsh conditions and parasites, featured on the ABC *Catalyst* program, show that these animals are not your average bovines.

The market and benefits:

- good-natured grass-eaters;
- hold condition in harsh environments;
- easy calving and longevity;
- reduced use of chemicals;
- good hybrid vigour;
- fullblood Boran (ancestors are Boran only);
- upgrade Boran (crossed from any breed).

Trying to describe their nature is like explaining or describing how to ride a bike – unless you experience it, you don't know. Visit a Boran breeder and let the cattle do the rest. For good measure, see the list of breed qualities on the Boran Cattle Breeders Society of South Africa website.

Bramalows

Bramalows are small beefy cattle derived from the Ausline and the Brahman. Bramalow breeders have carefully selected the superior traits of the Ausline and Brahman to develop an animal that maintains the great beef characteristics of Ausline but adds the hardiness of the Brahman, producing an animal that is suitable for smaller landholders in a range of climate types. Bramalow breeders are members of the Ausline Cattle Association and Bramalows are registered with the Ausline Cattle Association as percentage cattle.

The genetic makeup of the Bramalow allows the breeder to design animals to suit particular conditions. This means that Bramalows are extremely functional cattle. Animals with varying percentages of Brahman and Ausline genetics are accepted as Bramalows (75/25% to 25/75% Ausline/Brahman) This means that the Bramalow will always be attuned to its environment. Producers in more tropical or more demanding climates may opt for a higher Brahman content, while the higher Ausline content may suit other landholders.

Bramalows are sleek-coated, which makes them resistant to ticks and buffalo flies and untroubled by extremes of heat. They are also resistant to pinkeye. The females have very good udder structure and milk supply. They calve easily and

Bramalow bull.

usually are joined by 15 months of age. The calves grow quickly as a result of the females' good mothering ability. The bulls are very fertile and their size makes them ideal for use with first-calf heifers of other breeds. They produce very saleable, beefy calves with no trouble, thus allowing heifers to calve at two years of age with no difficulty.

Bramalows are naturally polled, but some animals will show evidence of scurrs (small, stubby horns detached from the skull). These are quite acceptable. Dehorning is unnecessary. Bramalows are mainly black in colour, but there is no exclusion of other colours. The influence of the Ausline will mean that the black colour is dominant but the inclusion of the Brahman genetics can result in variations. The height of the Bramalow is prescribed only for the show ring, where they will usually be about frame score 3; commercial animals may vary a little from this. They should always be moderate in size.

Bramalows will readily thrive in all parts of Australia, but their particular characteristics of tick, parasite and heat resistance and their size make them ideally suited to smaller farming operations in hotter areas. Their minimum maintenance requirements, polledness, beefiness, early finishing and longevity make them the first choice for farmers interested in raising easy-going commercially viable cattle.

Dexters

The Dexter is a naturally small dual-purpose breed of long standing which originated in Ireland, not a recently miniaturised version of a mainstream breed. The ideal height for bulls is between 102 cm (40") and 112 cm (44") and for cows 97 cm (38") and 107 cm (42"). They have great eye appeal and come in three colours – black, red and dun. Dexters are intelligent and docile, with an easy-going temperament and considerable character. They forage well and are light on the ground.

Modern beef and dairy breeds are very specialised; a dual-purpose breed should be a combination of both. In Ireland the Dexter was the smallholder's breed. It had to withstand fairly basic conditions and convert rough scrub and pasture into good-quality beef or milk. However, Dexters began to be imported into England towards the end of the 19th century and affluent landowners started to breed them for pleasure as park cattle. Dexter fortunes varied after World War II, numbers dropped alarmingly and the breed was rated as Endangered by the Rare Breeds Survival Trust in the 1970s. Fortunately a revival of interest has ensured that the Dexter is now bred in sufficient numbers across the world to be classified as a minority breed, no longer at risk.

Dexter cattle first came to Australia in the 1880s, with a second wave of imports a century later. Since that reintroduction the Dexter has established itself as a popular choice for enthusiasts living a relaxed lifestyle in rural residential areas where the smaller-size blocks are ideally suited to these cattle.

A Dexter cow can raise her own calf and also supply milk for the household or just graze and raise a good beefy calf, all the while keeping the fire hazard to a minimum. Dexter milk has smaller fat globules than that of mainstream dairy cows, giving a degree of natural homogenisation which contributes to the fine texture of Dexter butter, yoghurt and cheese. Organic or biodynamic Dexter beef is becoming more widely available from specialist outlets and is appreciated for its flavour and tenderness as well as the smaller portions that appeal to today's market.

Dexters respond well to a little extra attention and the true delight and enjoyment of owning a Dexter can only be fully appreciated after spending time with them.

Highland cattle

For centuries the Highland breed lived in the rugged, remote highlands in the west of Scotland, and the islands off the west coast of Scotland. The extremely harsh conditions that prevailed saw only the fittest and most adaptable animals survive to carry on the breed. At this point in time these animals were smaller and mostly black (known as the 'kyloe'), but as the centuries rolled on, the adaptability of the breed showed through and they now have a number of purposes and appear in numerous colours. In addition to the black colour, Highlands also come in red, dun, yellow, white, silver dun and brindle.

You will see folds (Scottish term for a herd) of Highland cattle in many different settings all over the world. In Australia, for example, they fare just as well in Queensland as they do in Tasmania. This flexibility of purpose stems from their

Highland bull.

hardiness (being rugged, somewhat disease-resistant and having very few calving problems) and the fact that they haven't changed much over many centuries of breeding. They are the oldest registered breed of cattle in the world, with the first herd book being started in Scotland in 1885. They are the preferred breed for conservation grazing because of their non-selective grazing (eating many species that other breeds will not touch), hardiness, modest size and longevity.

Primarily they are used for boutique beef production, because of their succulent and tender beef – the meat is well known for its marbling and flavour. Besides premium beef production, many Highland breeders market Highland hides as decorative floor rugs, and some even polish horn sets to sell. However, to many breeders on smaller properties, they are just an aesthetic pleasure to have in the paddock. Highland steers are especially suited to this use.

The popularity of this breed is unquestionable, because of the cattle's history, majestic appearance and, significantly, extremely docile temperament. So grand and majestic is the breed that the Queen of England has one of the finest folds in Scotland, at her summer retreat at Balmoral Castle.

Their beauty is apparent from the first time you lay eyes on this ancient breed from the Scottish Highlands. 'We haven't stuffed them up yet' were the words used by one highly regarded cattle judge.

Miniature Galloways, Miniature Belted Galloways, Miniature White Galloways

Galloways are a very old Scottish breed and are recognised as the oldest polled beef cattle in the world. The earliest recorded evidence of the breed in Australia is 1858. Miniature Galloways have all the characteristics and qualities of Galloways, only offering it in a smaller package which may be beneficial for small acreages.

For Galloways to be registered as Miniatures, they must comply with strict height requirements. Bulls cannot be more than 125 cm at mature age, while females cannot be more than 120 cm at mature age.

Galloways and Miniature Galloways, with their unique double coat of hair, can be black, dun or red in colour in each of the three types – Galloways, Belted Galloways and White Galloways.

Galloways are an extremely versatile breed that withstands extremes of temperature and climate. They have the ability to forage and thrive in marginal conditions or perform outstandingly on high-grade pasture. Being non-selective grazers, they are friendly to the environment and aid in pasture management.

Galloways are an extremely fertile breed, regularly producing a vigorous live calf. The Galloway cow is noted for ease of calving, is a protective mother and has an abundant supply of milk. Galloway bulls are noted for being prolific breeders.

Galloways are long-living, very resistant to disease and easy to manage. They create strong hybrid vigour due to the purity of the breed.

Miniature Belted Galloway cattle. Photo courtesy Wilkamdai Galloway Stud.

Galloways exhibit great carcass qualities, including excellent muscling, marbling and high-yielding saleable meat. Using Galloways and their crosses, it is easy to obtain optimum fat cover and muscling to suit most trade requirements.

Galloways have excelled in carcass competitions at most of the major events and have produced excellent results in grazing trials against all breeds. Galloway beef is renowned as being a tender, tasty and succulent beef of preference around the world. Galloways will turn feed into quality red meat economically and efficiently.

Miniature Herefords

The origin of Miniature Hereford cattle is in Herefordshire, England. The Miniature Herefords we know today are descendants of pure Hereford stock selectively bred since the 1970s. The trend at that time was 'bigger is better' but one breeder went against the trend and selectively bred for temperament, hardiness, meat quality and feed conversion.

The Miniature Hereford is best described as a small chunky version of the larger Herefords. In contrast to the tall long-legged larger type, Miniature Herefords are thick, deep bodied, short-legged, muscular and chunky.

Miniature Herefords are very hardy and adapt well to various climates. Currently, Miniature Herefords are bred in all parts of Australia. They have no special needs other than what is required for the care of regular cattle.

Miniature Hereford calf.

Why Miniature Herefords? Advantages of the breed are:

- investment: the herd numbers of these cattle are still very low. This allows great potential for investors;
- quality meat: the quality and yield of the Miniature Hereford is excellent. This makes minis a commercial reality;
- easy to handle: their size and placid nature makes Miniature Herefords ideal, even for those who have no previous experience with cattle;
- attractiveness: to many people the Hereford is a most appealing breed purely for its looks. Miniature Herefords are striking to the eye;
- carrying capacity: with the potential to carry more head to the hectare, Miniature Herefords are an ideal alternative for small and large farmers alike.

It is nearly 20 years since the first importations of embryos and a small number of adult minis from Canada. One of those original imported cows is still very much alive – now aged 19 (as is her daughter, born in quarantine 17 years ago) and in excellent health – a testament to the longevity of the breed. The Australian herd continues to grow, with constant enquiries via the Australian Miniature Hereford Cattle Association re the availability of animals for sale. Recent importations of both polled and horned semen have enabled the expansion of different breeding lines, and ensure the viability of what has become a very popular and successful breed since its introduction to Australia in 1997.

Miniature Herefords are registered with Herefords Australia as purebred Herefords and certified by the Australian Miniature Hereford Cattle Association Inc., which was established shortly after the first importations to promote and protect the integrity of the breed in Australia.

Nadudana cattle

Nadudana (pronounced Nar-Dar-Nah) are a naturally small *Bos indicus* breed from Ceylon, now known as Sri Lanka. They are one of the few naturally miniaturised breeds and one of the most ancient breeds in the world, dating back to 3000 BCE. The name means 'small cattle' in Hindi and they were first imported to Australia from North America in 1995. They are considered to be globally critically rare.

These diminutive animals are physically similar to their larger cousins the Zebu cattle, i.e. Brahman, and are properly proportioned with the exception of their size. The mature Nadudana animal should not exceed 107 cm as measured across the back, behind the hump. Nadudana cattle are very family-oriented. Easy calving, most cows will calve very close to the herd. Many calves are born white, with a reddish 'cap' on top of the head, some are born red but change to black, grey or white by adulthood while others may stay red.

Breed advantages:

- general appearance: a robust, stylish, dual-purpose animal, active and vigorous, showing breed character and vitality, well proportioned and muscled. Bulls, masculine and cows, feminine;
- temperament: quiet, docile and generally alert. Many owners have fallen in love with their quirky natures. This breed has a wonderful sense of fun, which is lacking in many other breeds. The cattle actually play together, which is a constant source of delight. They have personality!;
- efficiency: intake of higher levels of low-quality feeds, high meat to bone ratios, higher feed conversion ratios than full-size cattle;
- heat tolerance: dark pigmented eyes and muzzles reduce risk of cancer, sleek short-haired coat with increased number of very efficient sweat glands, lower metabolic rates generate less heat. The only small *Bos indicus* (Zebu) tropical breed in Australia;
- parasite resistance: a major economic consideration as fewer chemicals are used. Sleek coats with sweat glands do not favour attachment of tick larvae and repel flies and the breed is resistant to internal parasites;

Nadudana cow.

- beef: an evolving market for quality small 'boutique beef' cuts exists within Australia. Quality beefy, dairy cross calves;
- small farms: suited to small farms and on average can be stocked at 2.7 beasts to each standard size animal, which allows commercially viable stocking rates with progeny finished on grass. Often used as pets (lawn mowers), so the human side is quite influential on the breed decision;
- adaptable: Nadudana cattle can adapt to climatic and environmental extremes.

Nadudana Cattle Queensland Inc. has to date maintained an open book breeding policy, which allows breeders to grade up using registered Nadudana bulls over non-Nadudana cows and Nadudana crossbred cows. The Association has breeders located in all Australian states and in New Zealand, and its herd book contains more than 500 registered animals.

Square Meaters

Square Meaters are an Australian breed of beef cattle, recognised as a separate breed with its own herd book in 1996. Breeders are predominantly in Australia, in all states except the Northern Territory, and there are breeders in New Zealand and the US. Square Meaters were developed to produce a finished domestic carcass within 12 months. These compact cattle are **neither small nor miniature**, but are nevertheless ideal for smaller acreage.

Square Meaters are a measured breed; females must measure between 100 cm and 110 cm at the shoulder, bulls between 103 cm and 113 cm at 12 months of age.

These regulations are designed to provide a safeguard against animals becoming too small or regressing to a taller, slower-growing, later-maturing type. Bulls weigh around 800 kg at maturity, females around 500 kg. Typically, Square Meaters have a deep well-muscled body on short legs (see photo in colour section).

Square Meaters were developed by selection, from high-performance and early-maturing Murray Grey cows and bulls, and have the following characteristics:

- a quiet disposition – they are easy to handle, easy on fences and quiet in the yards;
- are naturally polled;
- have a medium frame and early-maturing growth;
- are heavily muscled with even fat cover;
- have good fertility and mothering ability: females can produce their first calf at two years of age and then produce a calf every year. A good supply of milk makes for a healthy, quick-growing calf;
- have low birth weights: calves typically weigh 20–30 kg and are born easily. The owner can sleep easy at calving time;
- have good 'doing ability' even in poorer conditions: their compact size means they hold their condition longer when times get tough and larger breed females are starting to fade.

Square Meaters are recognised for their beautiful and characteristic colours, which range from silver through grey to a dark grey, which includes brown shades to a dark chocolate colour. The colours are solid, i.e. they don't have patches of lighter colour. The skin has a dark pigment, which provides some protection from sunburn and skin cancer. These are cattle you'll be proud to have in your front paddock.

Square Meater breeders are friendly, helpful and enthusiastic. They welcome new breeders and are keen to help newcomers. The Square Meaters Cattle Association offers truly personal service to all its members, and aims to keep breeders' costs as low as possible.

Q. Is the Lowline breed derived in part from the Dexter cattle, these being also short of stature, with pleasant disposition?

A. The Australian Lowline is not derived from the Dexter. Dexters are a very old breed originating from Ireland and Lowlines are an Australian breed derived from Angus. They are similar in size, with the Dexter being marketed as dual-purpose breed and Lowline as a beef breed.

Q. Do all small breeds carry the dwarfism gene?

A. The only small breed that I am aware of that carries the achondroplastic dwarfism gene is the Dexter.

Q. How feasible is it to have a miniature female cow as a pet and for milking?

A. Small breeds are perfect for this scenario. Their docility means they make lovely pets and those that are marketed as a milking breed combine the two characteristics perfectly.

Everybody has a different reason for wanting to get involved with cattle. Here are just a few examples.

I am looking into buying a small acreage (seeking a 'tree change') and am interested in learning more about small cattle for my wife and I to farm.

Basically I am really keen to find out more information on small cattle and am looking at the possibility of purchasing some. A bit of background for you: I grew up with cattle, my parents and grandparents had properties where we predominantly had Brahmans. However, I am now married with two small children and we live on a cane farm in northern Queensland. We have 3–4 hectares available with access to a creek that we would like to make use of. I loved growing up with cattle and would love my kids to have a similar experience.

I live on 20 acres east of a small mining town in Queensland. After living here for the past 12 months I am in the position to fulfil a lifelong dream of owning and breeding cattle. I have researched a number of breeds (although I'm far from an expert). At this point in time I just want to start up as a hobby and hopefully develop into something bigger when I can devote more time to it. Toward this goal I see no point in purchasing inferior bred animals. As I do not know a lot about breeding, what would be a good starting point to develop a strong lineage?

My husband and I have a small property (40 acres). We are seeking information about small cattle with the thought of starting a small stud, which we would like to run now and into our retirement years. We currently run steers, but have in the past had a Charolais stud.

Ausline bull.

Ausline calf.

Australian Lowline female.

Bramalow heifer.

Dexter bull and cow.

Dexter bull.

Highland cow.

Miniature Belted Galloway cattle.

Miniature Galloway.

Miniature Hereford cows and calves.

Miniature Hereford bull.

Nadudana bull.

Square Meater cows.

Height comparison of a small breed with a larger breed. Both these Grand Show Champion British breed bulls are the same age, i.e. 22 months old. The Red Angus bull (back, GK Red Absolute Power) measures 153 cm at the hip and weighs 970 kg while the Dexter bull (front, Bircham Lord Litcham) measures 114 cm and weighs 540 kg.

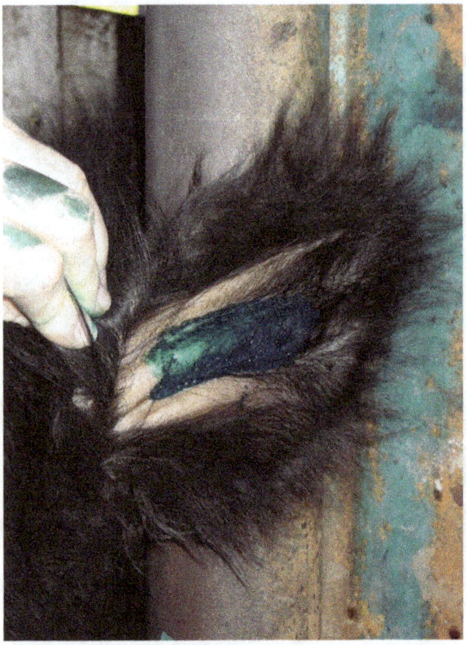

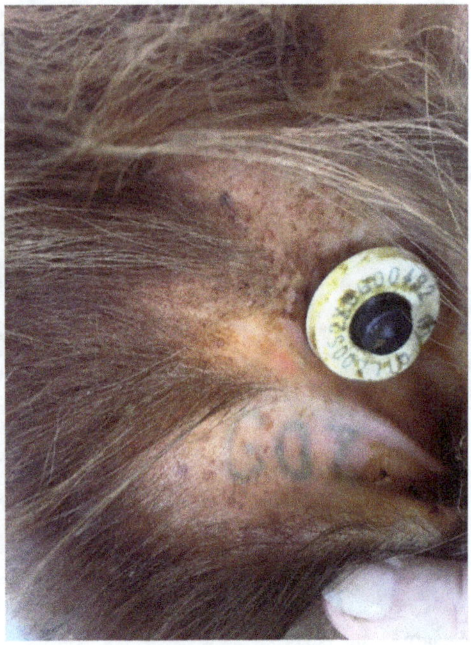

Left: The ear is cleaned and the central area to be tattooed has had paste applied. Right: The tattoo is visible in the bottom section. The NLIS device is also visible.

A Dexter Field Day with kids of all ages enjoying the fun.

5

Selecting stock

My advice when it comes to purchasing cattle is to buy the best you can afford and take extra care in choosing your females. When I first started I bought two outstanding quality females to form the core of my breeding herd. It is a decision I have never regretted – those initial quality genetics are still evident in my show-winning stock today. You can always change a bull if he is not producing the progeny you want but it can be a very expensive process to change your female herd. This doesn't mean that females are more important than males, it is just that only one bull is required for every 30–40 females so your initial capital outlay is far greater in purchasing your breeding females.

If you are buying stud stock then check with the relevant breed society, as some produce 'buyer beware' information that you will find useful. For example, some associations require that all animals be DNA-typed and parent-verified, so ask the vendor to make sure this has been done. Some also state that registration of animals cannot be transferred to the new owner unless the vendor is a current financial member of that association. Verify the animal's registration papers, make sure it is tattooed correctly and establish who is responsible for transfer fees prior to purchase.

Choosing your breeder

It is a good idea to check the credentials of the breeder of your livestock. This is particularly relevant and important if you are buying stud stock. Obviously if you are just after some nice quiet grass-eaters then the following list is not so significant. Here are some attributes that may be considered.

- Vision: a successful breeder needs to have a very clear and defined vision for the future and know what strategies they will use to achieve it. Find out their plans, where do they see themselves in 10 years, what new markets do they hope to access?
- Experience: it takes a long time to put a good herd together and be able to make meaningful genetic progress. What genetics traits do they consider important? When you view their herd, can you see what they are aiming for and is it evident in their breeding program?
- Management: an organised, efficient system also acts as a guarantee. Ask to see the animal's vaccination and joining records.
- Client focus: a product guarantee and after-sales service that is ongoing and educational. Is the breeder willing to help with your enquiries? Will they be there for you when you need to ask questions? Will they mentor you if need be?
- Integrity: is the breeder honest and trustworthy? Will they sell you what you want, not what they want to sell you? If you want show quality, make sure you get show quality.
- Technical knowledge: the object of breeding is to improve and build on your genetics with each generation, so you may need assistance or advice for improvement of your herd. Is the breeder aware of the latest scientific developments and do they encourage the use of these performance indicators? Beware of dinosaurs!
- You pay for what you get, like most things in life. It is economic lunacy to buy a bull for $500 or $700 because it is cheap. If its breeder has no vision, no guarantee and is not interested in looking after you, then you are buying yourself trouble. A breeder with all the attributes listed above may be charging $2000–3000 for a bull, which you may think is ridiculously expensive. But in comparison it may be the cheapest investment you have ever made.

Desirable traits

When choosing your animals, consider these factors.

- Make sure the animals you purchase conform to the breed standards. The main aim for stud breeders to is to have an even line of cattle that are similar in type.
- Make sure your animals are structurally correct.
- Make sure your animal's physical appearance is gender-specific and balanced. That is, that all traits are in proportion – you are protecting against single trait selection. For example, in the dairy industry the emphasis is on high milk

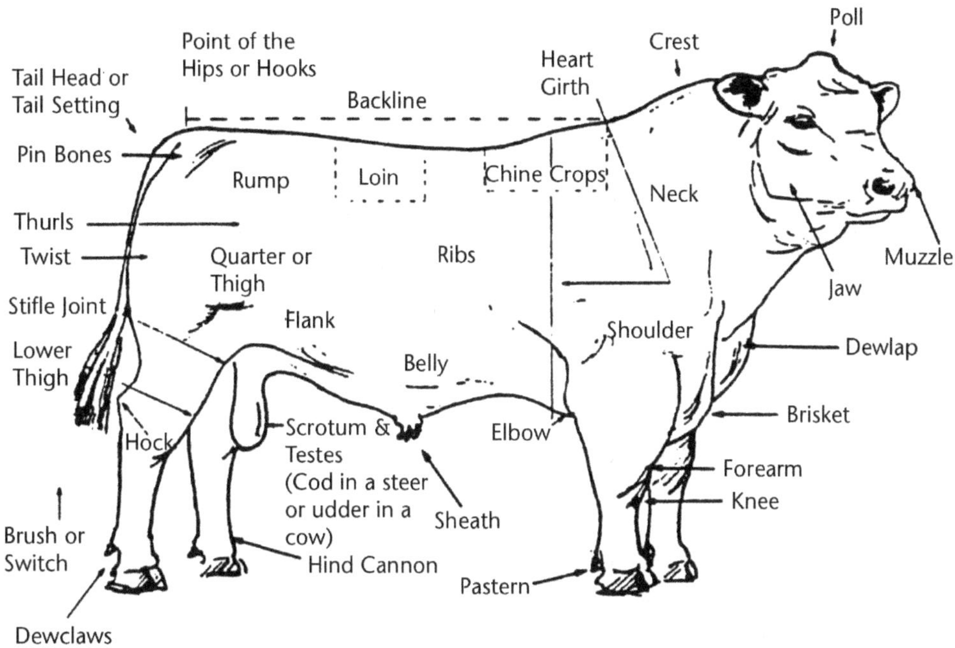

Parts of a beef animal. Source: Beef Cattle Judging.

production and for many that is the only selection trait considered. This has resulted in high-maintenance cows with incredible nutritional requirements, calving difficulties and low conception rates. Similarly, those within the beef industry chasing huge frame scores have produced heifers that are unable to calve naturally. This results in damaged heifers and/or dead calves, and enormous economic loss.

What you are looking for is balance and eye appeal. This is determined by the proportions of the animal and how well they fit together.

Females

The female's role is to produce offspring each and every year. She must be able to give birth to a live healthy calf and produce enough quality milk that her calf can grow and survive. The cow must be able to maintain her body condition so that she can continue to reproduce and graze efficiently in competition with other cows in the herd.

- Femininity: in the show ring you will often hear the judge talk about femininity in females, meaning they want a cow to be pretty, graceful and feminine. Her neck and shoulders should be slender and well in proportion. Wide heavy shoulders make her look masculine and narrow pinched shoulders often indicate a high-maintenance female.

- Wide across the pins: long from the hips to the pins and wide between the pins. These 'child-bearing hips' are a feminine trait and indicate high fertility, allowing adequate room for a calf to be expelled during birth.
- Tails: thick and long tails indicate low-maintenance animals.
- Udder soundness: there are four separate quarters that are not connected. They should all be even and not pendulous. The ability to make milk is one of the cow's most important functions so you need all four quarters working properly. Teats should be small sized and well placed. Big bottle teats are not desirable as young calves sometimes find it impossible to drink from them.
- Volume: females that have deep girths and deep flanks indicate low-maintenance, calm animals. It also means that they have a greater beef carrying capacity.
- Head: it should be feminine looking and not bullish. Nostrils should be wide apart and eyes should be hooded, meaning they are well into the head and not 'popping'.
- Skin type: shiny hair results from the natural secretion of body oils and is indicative of a healthy animal. It makes the animal more parasite-resistant. Dull hair indicates health problems. Loose skin indicates a relaxed animal.
- History: has she had a calf every year? If not, find out why not. You need fertile animals that return to calf easily.

Males

The bull's role is to breed and pass on quality genes for low-maintenance, high-fertility animals. You want a well-balanced masculine bull that is fertile and has a high libido. You need a sire that is capable of walking the walk and talking the talk without excessive fighting. Fighting excessively to prove strength is the sign of a subfertile bull.

- Appearance: they must look like a bull and be impressive in appearance. They should take your eye the minute you see them. Animals that are deficient in male hormones will be more feminine-looking.
- Head: they need to have a strong masculine head. You want a wide strong jaw, broad nostrils and well-set alert ears with well-hooded gentle eyes.
- Neck: bulls' necks should be short and muscular, unlike females whose necks should be long and feminine. A short muscular neck is a good indication that the bull will have a healthy libido.
- Testicles: need to be formed correctly, i.e. even in size, hanging together and not twisted. The old saying is that they should look like two beer cans hanging in a sock! Bulls with bigger testicles produce more sperm, giving a higher chance of conception, and also produce more fertile females. Testicles that are too close to the body can result in overheating and 'fry' the semen in hot

weather, testicles that hang too low are more likely to be injured. All bulls have teats on their testicles. Make sure the teats are not developed, as enlarged teats are a sign of hormonal imbalance.

- Sheath: should not be too pendulous as this can result in injuries and infections. Some breeds are more susceptible to these problems and buyers of those cattle should be critical in their selection.
- Skin type: shiny hair results from the natural secretion of body oils and is indicative of a healthy animal. It makes the animal more parasite-resistant. Long hair indicates health problems. Loose skin indicates a relaxed animal.
- Tracking: the bull should place its hind feet in the prints of the front feet as it walks freely. It should have a free-moving gait.
- Muscle pattern: you want bulls to have a good even muscle pattern with meat on their hindquarter. You want their hindquarter to be in proportion with their shoulders. Don't buy bulls that are big in the shoulder and head and wasted in the butt.

Structural soundness

The 'structure' of an animal is the manner in which the skeleton is formed. You should purchase cattle that are structurally correct, as many of these traits are highly heritable. Not every animal will be show quality but good structure is important in all types of breeding animals, both stud and commercial. Structural faults can be costly as they can reduce the working life, libido or fertility. Correct structure will enable the animal to walk, feed and breed to its maximum ability.

Feet

If your animals have bad feet then it is highly probable that their progeny will as well. Pay special attention to their legs and feet and avoid overgrown, curved or scissor claw hooves or those with an abnormal walk. Overgrown uneven claws are usually indications of bad limb structure or early signs of hip arthritis. Likewise, very short feet are often associated with overstraight legs. If an animal does have bad feet you need to take into consideration whether it is actually a structural fault or a result of its environment. Cattle that are raised on very soft country, e.g. rich black soils, often grow rather long claws but this is a direct result of their environment rather than a structural fault. Rocks and rough terrain are a great way to keep their toes short.

Legs

Cattle that have legs that are set incorrectly will have more problems, especially under stress such as hilly country or when breeding. If the animal is too heavy for its bone size this will cause a sickle hock condition causing strain on the muscles

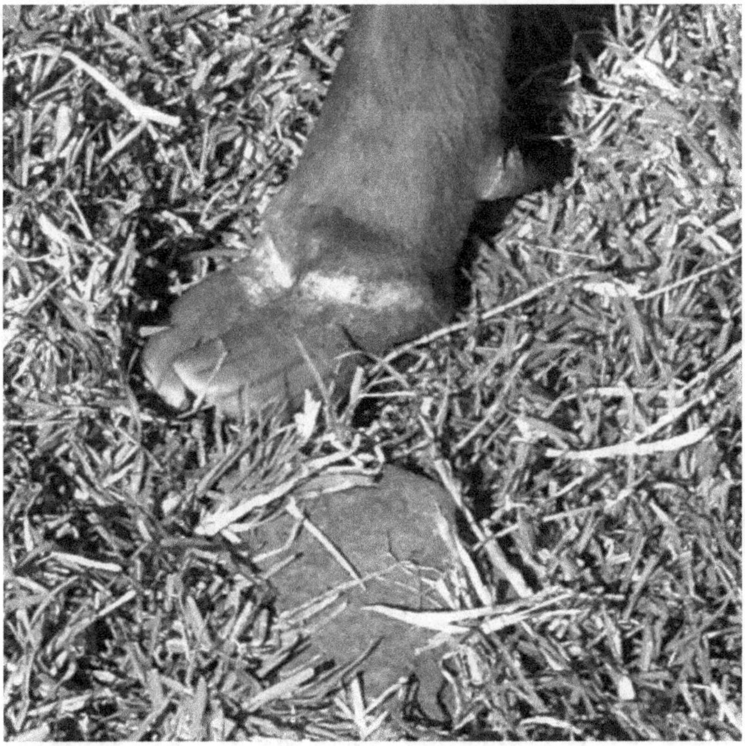

A photo of very bad feet.

and bones of the stifle joint, hips and legs. Sickle hocks occur when the joints are over-angled, leading to overstepping and long overgrown claws. Mild cases are not a concern but severe cases can cause strained ligaments, damage the dewclaw and affect a bull's ability to serve females.

Post-legged animals, or straight hocks, will normally walk short and put more pressure on the stifle joint. This can result in torn ligaments, tendon and muscle problems. Cattle that don't have the flexing and shock-absorbing effect of the structurally sound animal are prone to severe wearing of the hip joint, leading to arthritis. Bulls showing arthritic problems are reluctant to serve many cows, as the condition can be quite painful. Penis damage can occur due to serving accidents, causing the libido to decline.

When a bull mounts a cow, he straightens the joints in his hind leg. When he thrusts, he further straightens the leg, placing enormous stress on all joints, but particularly the hock. If these joints don't have enough angulation, they become swollen and painful, eventually leading to their breakdown and the bull's reluctance to serve the cows.

The majority of the feet and leg faults are genetic but they can sometimes be caused by environment. Excessive overfeeding of grain, mineral deficiencies and soft soil can also contribute to these symptoms.

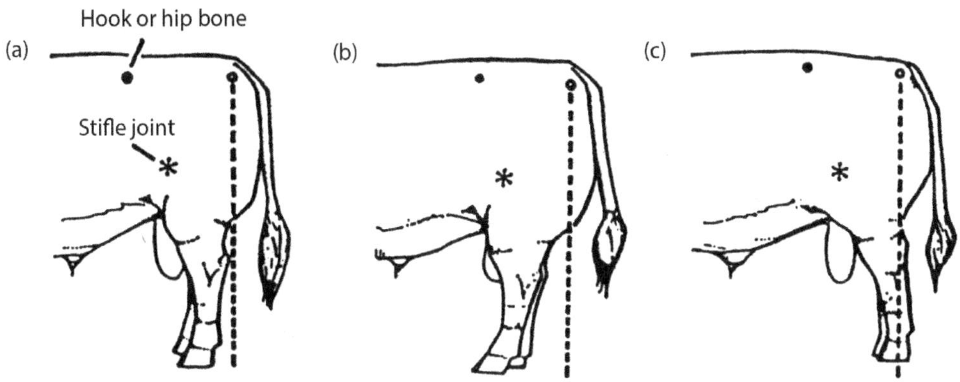

Examples of leg settings. (a) Correct leg set. There is a straight line from the pin bone to the hock to the dewclaw. (b) Sickle hocked animal. The hind legs are too far under. (c) Post-legged animal. The hind legs are too far back.

Walk

Look for a free-moving gait. The hind feet should step into the footprints of the front feet; in the show ring this is referred to as 'tracking'. Over- or understepping are indications of structural problems, as are uneven footprints of the claw. It is not uncommon for the small breeds, where you have an exceptionally long animal, to understep, as it is just not physically possible for them to track evenly.

Shoulders

You are looking for an animal with nice smooth shoulders. An angle of 45–60° is considered acceptable. An animal whose shoulder blade is tipped forward (i.e. straight) has less angle at the shoulder joint and elbow joint, which reduces the shock-absorbing ability. A straight-shouldered bull tends to walk with a short uneven gait. He may even have difficulties raising his head much above his backline and will walk with his head carried low. Straight shoulders often correlate with straight hind legs.

Mouth and jaw

The muzzle should be wide for efficient grazing. The teeth on the lower jaw should meet squarely with the upper pad. Animals with overshot jaws (lower jaw protruding) and undershot jaws may have difficulty grazing on pasture, especially when there is a shortage of pasture.

Loin and back

Often in the show ring you will hear a judge refer to an animal's 'strength of spine', meaning a straight and wide backline. The judge is looking for a strong straight spinal column all the way from the hipbones to the top of the shoulder. The width in a bull's top line will help indicate the degree of muscling. Heavily muscled bulls

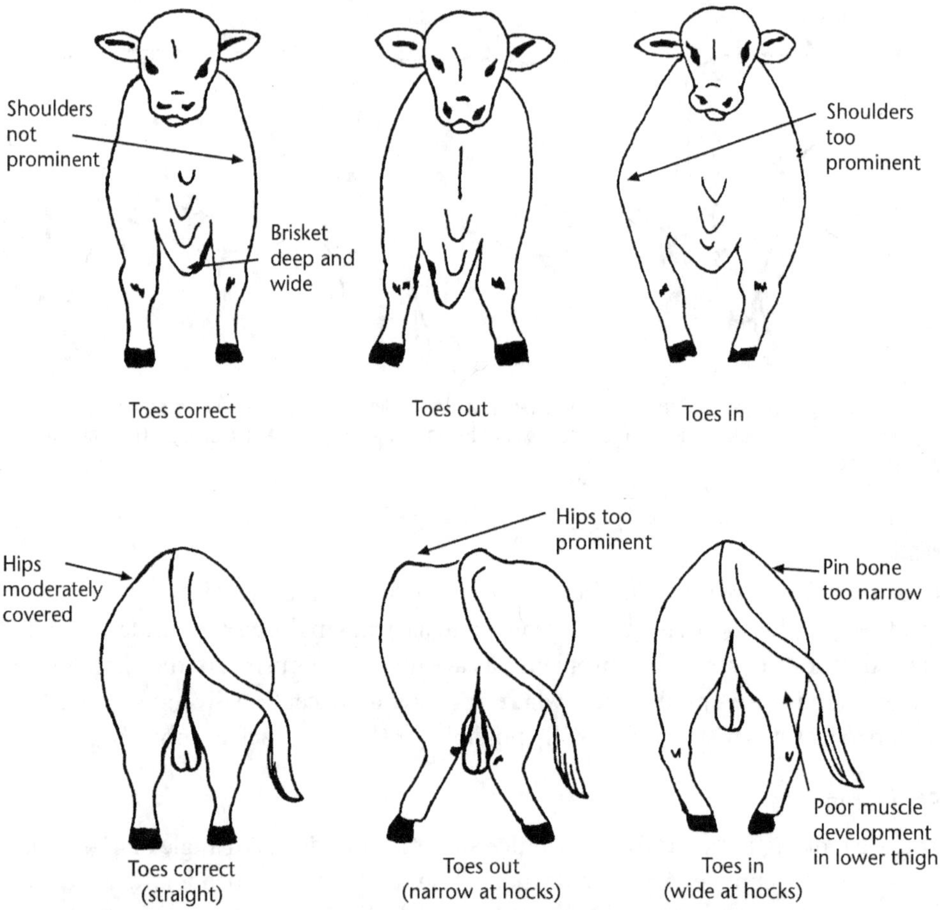

Bone structure and muscle. Desirable features are shown on the left and undesirable features are shown in the middle and on the right.

will be wide and bulging in their top line, especially behind the shoulder, and in the loin. Weakness in the spine just before the hip will cause birth difficulties. This defect, combined with high pinbones, may compound the calving problem so females with a weak, flat or broken loin are considered to be lower in fertility.

Udder

The udder is one of the most important aspects of the females as udder soundness affects milk production, milk composition and calf weaning weights, the incidence of calf mortalities and the duration the cow stays in the herd. All these factors affect the overall profitability and productivity of your enterprise. Cows with medium-sized well-attached udders wean faster-growing and healthier calves and are able to remain in your breeding herd for longer.

The ideal udder is snugly attached, symmetrical and of moderate length. An udder is divided into four quarters, which should all be evenly balanced with

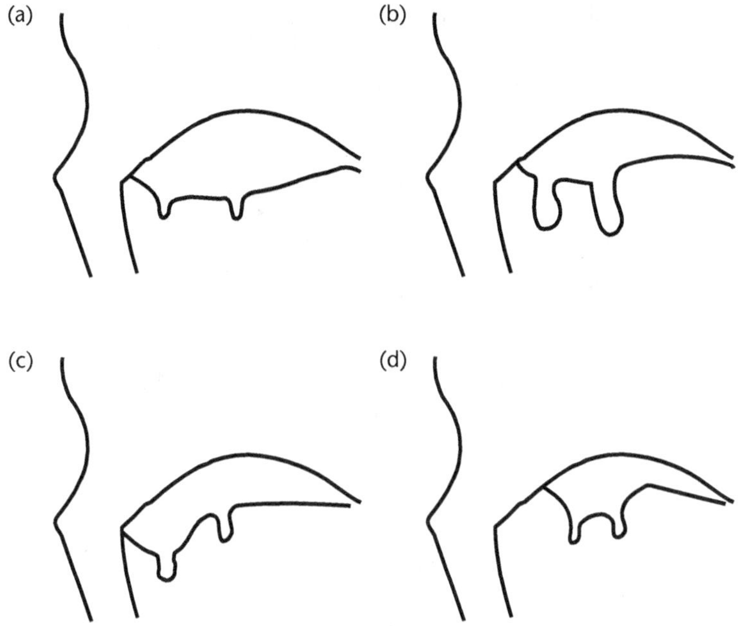

Examples of udder shape. (a) Nice even udder with good capacity for milk and well-balanced and spaced teats. (b) Bottle or enlarged teats, making it difficult for calves to drink. This can result in mortalities. (c) Less capacity in the forequarter, which often occurs with age. Difficult for calf to drink from the front. (d) Poor capacity for milk production and undesirable shape.

medium-sized teats placed squarely under each quarter. A side view of the udder should show a level udder floor without quartering.

Avoid large bottle-shaped teats and excessively long thick teats, as they cause the greatest problems to newborn calves. Balloon, bottle or funnel-shaped teats make it difficult for the postnatal calf to suck, increasing the incidence of ill thrift and calf mortalities. This is a highly heritable trait and cows with these defects should be culled.

If the udder is too pendulous then the newborn calf often finds it difficult to suck. The fore udder should be of moderate length and strongly attached. Extra-long fore udders are 'meaty' and frequently break away from the body wall. The rear udder should be attached high to the body with moderate width and show a strong median suspensory attachment. The udder texture should be soft, pliable and free from congestion and hardness.

Bull reproduction soundness

Most small breeds have exceptionally large testicles for their size. The minimum size for a standard British breed bull is 32 cm and most small breeds meet this criterion easily. The bigger the better: size affects the fertility of the bull's daughter. Bulls with bigger testes sire daughters with better fertility. *Bos indicus* breeds have slightly longer, 'smaller diameter' testicles. Scrotal size is not the only criterion when

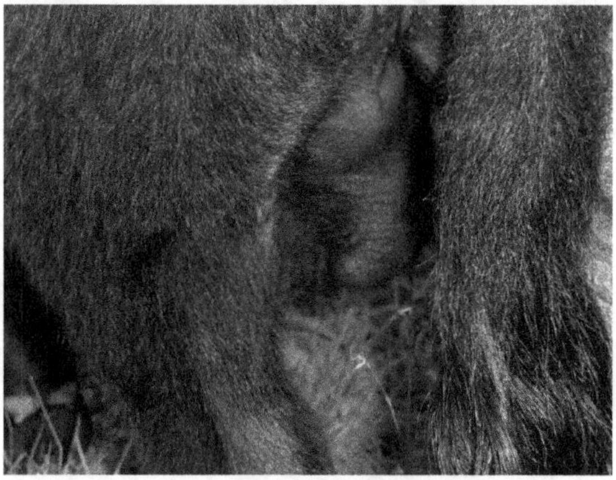

Unevenly developed testicles.

selecting a bull. The testes need to be 'firm and springy' and free of lumps. Soft testes may indicate an issue in the semen quality. Make sure they are even in size, as one small testicle or two small testicles are the most common abnormalities found in *Bos indicus* bulls. The problem is heritable and associated with reduced fertility.

The sheath should be trim and close to the body to avoid injury and/or infection from grass seeds. The prepuce (the fold of skin covering the penis) should not be slack. A bull that lets his prepuce hang out for long periods is undesirable. You will often see a handler check their bull in the judging ring to make sure the prepuce is not slack and hanging down. If it is, the handler will quickly give it a tap with the show cane to make it retract – hopefully before the judge sees it!

A client sent me a photo with the following message. *He has developed a problem, and the older he gets (now eight months) the more noticeable it is. His little-boy bits are not developing evenly. One side is longer than the other. Both testes are down. He would still breed, but looks funny. He is too good to steer.*

I totally disagree with the last comment. He could easily produce calves that have the same fault and any animal that has poorly developed testicles needs to be castrated. Animals like this will do no justice to your breeding program. They should not be allowed to reproduce.

Q. We understand that a bull's stud life is around five to six years, but what about a cow?

A. Cows can live and reproduce until they are 17 years of age and older. Often their ability to produce milk will give out as their udders break down, as will their teeth, long before they are unable to conceive. There is no reason why a bull won't last longer than six years if he is well looked after. I have a nine-year-old bull that still works exceptionally well.

Other factors

Temperament

Temperament is high on my agenda of importance, especially for those with small acreage. Don't tolerate poor temperament, as it appears to be highly heritable. It reduces cattle's ability to perform as they stress badly when yarded and handled, as will their progeny, they will have a higher incidence of bruising on carcasses and often have dark-cutting high pH meat. Excuses should not be made for cattle with temperament problems. You will regret it later!

When you visit your place of purchase or various studs, make sure the cattle are quiet so that you can walk easily among the animals without them taking off to the other side of the paddock. Exceptionally quiet herds will not be perturbed by your arrival and often don't even bother to get up. Flighty or nervy animals will react immediately. Quiet cattle will breed quiet cattle.

Length

I like my cattle to have lots of body length. Extra length produces more beef … a few extra T-bones for no extra cost! I like to produce long stretchy bulls.

Take your time

Visit a number of different breeders to see what type of cattle they are producing. Go to shows and watch the judging. See the type of animals that are winning and try and get an 'eye' for selecting animals.

6

Selecting genetics

You need to consider some other factors in determining the right animal for your herd. If an undesirable gene is introduced into your herd and you notice it within two years, it will take around 25 years to breed it out! Good genetics, once introduced into your herd, work like compounding interest at a bank. Each year your interest deposit will grow more valuable.

Maturity patterns

If you are wishing to enter the beef market, then it is advisable to understand the maturity patterns of the different breeds. The maturity of an animal is determined by the age at which growth ceases and fat is laid down on the carcass. Smaller breeds, on the whole, tend to be more early-maturing, i.e. they stop growing at a younger age and are able to put their energy into laying down fat and maturing rather than building bone. Some genetic lines mature faster than others. The best way to observe these patterns is by viewing an age class of young bulls in the show ring. The more early-maturing types will have a larger and more developed crest on their neck and will have started to fill out into their frame more.

Some breeds have characteristics better suited for certain feed conditions or particular environments than others: no one breed is the best for all environments and all markets. For example, European breeds are generally larger and produce more muscle and leaner carcasses than British breeds, which are renowned for their ability to fatten on less feed, are earlier-maturing and produce quality

Fact: Limousin have a natural muscling due to the unique F94L muscling gene. It is claimed that this gene protects against overfatness, particularly in heifers. Another gene, recently discovered by researchers at the University of Adelaide, is a modification of the F94L and appears almost exclusively in Limousin. It can increase the weight of prime cuts by 19% and overall beef yield by 7%.

marbled beef. The choice of breeds used in a crossbreeding program should take into consideration the complementarities of different breeds.

In my crossbreeding program I have had great success with using Ausline bulls over commercial cows that comprise 50% Limousin and 50% Murray Grey. I have won five royal show Carcass competitions with this particular cross and have twice taken out the Grand Champion Carcass at the Royal Queensland Show. This is an open competition against all breeds and comprises about 350 entries. The Limousin provides the muscle and growth while the Ausline and Murray Grey reduce the maturity pattern and provide the fat cover required.

Breeding systems

The aim of any producer is to strive for genetic improvement in their herd. This is applicable to both stud and commercial producers who are trying to produce the most efficient, quality animals for their target market.

Purebreeding: also referred to as straightbreeding, involves joining (usually unrelated) cattle from the same breed. I think it is extremely important to retain as broad a genetic base as possible. We aim not to include an animal more than once in a three-generation pedigree.

Line breeding: this is often referred to as a method wherein you keep the influence of any given ancestor to no more than 50%. Half-brother to half-sister is 50% influence. Stud producers often use line breeding when they want to reproduce a particular genetic quality in their herd, but beware of also increasing deleterious traits. Many small breeds have been developed using this method.

Inbreeding: this is where the influence of any given ancestor is greater than 50%. Father to daughter, son to mother, grandfather to granddaughter, grandson to grandmother etc. should be avoided.

Crossbreeding: animals of two or more different breeds are joined. Hybrid vigour is achieved through continuous crossbreeding and can be used to increase

You are sure to come across the old saying about joining closely related animals: 'If it works it is line breeding, if it doesn't work it is called inbreeding.'

the weight of weaned calves by up to 20%. Breed differences should be used to optimise production and carcass quality to meet the required target markets.

Composites: a stabilised mix of two or more breeds, e.g. Droughtmasters, Belmont Reds.

Grading up: this is a form of breeding in which each generation will be higher in purity than the previous generation. A number of small breeds allow this process. Animals produced by this method usually do not breed as true to type.

1st cross will result in a 1/2 bred animal
2nd cross will result in 3/4 bred animal
3rd cross will result in a 7/8 bred animal
4th cross will result in 15/16 bred animal that can also be classed, in some breeds, as a purebred.

For example:

An Ausline bull is joined to a Hereford cow	Progeny = 1/2 bred animal (i.e. 50% Ausline)
That progeny is then joined back to Ausline bull	Progeny = 3/4 bred animal (i.e. 75% Ausline)
That progeny is then joined back to Ausline bull	Progeny = 7/8 bred animal (i.e. 87% Ausline)
That progeny is then joined back to an Ausline bull	Progeny = 15/16 bred animal (i.e. 94% Ausline) or purebred animal

Hybrid vigour

Hybrid vigour (heterosis) is the increased performance associated with the combination of parents from two unrelated breeds. It is most evident when the breeds joined are genetically the least related, such as *Bos indicus/Bos taurus* cross. The new combination of genetic material can lead to advantages over and above the average of the two parent breeds or strains. Be careful, though: for this to be of economic advantage, the new production levels need to be above those of either parent strain or breed otherwise you are better off sticking with the superior parent line (see Table 6.1).

The increased performance traits are usually most noticeable in growth, fertility and survival rates. By using the right combination, increased growth rates in excess of 20% have been achieved and fertility traits are likely to be double those which might be obtained for growth. True hybrid vigour is when the crossbred animal is a terminal cross, meaning that both parents must be purebred.

Heritable traits

Heritability is the degree to which an animal will transmit a particular trait to its offspring. It measures the relative importance of hereditary and environmental influences on the development of specific quantitative traits. The higher the

Table 6.1: Percentage of maximum heterosis expected in progeny

Mating system	Maximum heterosis retained		Superiority over parent breeds	
	Individual (%)	Maternal (%)	Increased weight of calf weight weaned/ cow exposed (%)	Increased value of calf weight weaned/cow exposed at $1.30/ kg live weight gain ($)
2 breed cross				
A × B	100	0	8.5	16.5
3 breed cross				
(A × B) × C	100	100	23.3	45.0
Rotational crosses*				
2 breed	67	67	12.7	25.0
3 breed	86	86	20.9	40.0
4 breed	93	93	21.7	42.0
Composite crossbreed				
2 breed	50	50	11.6	22.5
3 breed	67	67	15.6	30.5
4 breed	75	75	17.5	34.0
5 breed	80	80	18.6	36.0
6 breed	83	83	19.3	37.5

*This breeding system refers to Bos taurus × Bos taurus or Sanga × Bos taurus crosses after about seven different matings.
Source: futurebeef.com.au/topics/breeding-and-genetics/crossbreeding-systems-for-beef-cattle.

Hybrid vigour: purebred versus commercial. These two steers are the same age, both 20 months old and born 12 days apart. The purebred Lowline steer on the left weighs 324 kg. The steer on the right is a 1st cross (50% Lowline/50% Angus) and weighs 498 kg.

percentage, the greater the chance that trait will be passed on to the progeny. The list in Table 6.2 is not complete but it gives you an idea of the percentage of heritability of some traits.

There has been far greater emphasis on genetic improvement programs for growth and carcass traits than on reproductive traits. Female fertility traits in general have a low heritability.

In Australia the larger breeds use EBVs (Estimated Breeding Values) to help estimate progress and setbacks for a range of production traits that you can expect from different matings. Small breeds are unable to use EBVs, as large numbers of animals of the same sex, age and production method need to occur for reliable data to be obtained.

Scientific advances

With the introduction of DNA and SNP genotyping for many selected traits and inherited disorders, science now assists us to produce better-quality cattle and beef. The nice part about DNA testing is that it requires only a simple hair sample with follicles attached: this certainly beats the old days of blood sampling.

At Vitulus we avail ourselves of a number of these products. Perhaps the most significant has been the colour coat test which allowed us to develop a whole new breed, initially called RedLine and now called Ausline, based on the ability to DNA test for the red and wild allele. DNA allowed us to accurately select and breed red animals knowing that we were joining known carriers of a recessive gene.

DNA typing

DNA can be used for determining qualitative traits, for parent verification and to test for inherited disorders. Qualitative traits are those that show a clear distinction between genotypes such as coat colour and horns. These genomic tests have the potential to allow producers to more accurately predict genetic merit for their herd. Once again, be careful of placing undue emphasis on single production traits and remember that environmental factors also affect carcass traits. As the technology improves, these tests will become better predictors of genetic merit.

- Parentage verification (e.g. that required for breed society registrations) or multi-sire mating analysis to determine parentage in extensively managed systems. The advantage of this is that breed societies which use this system can guarantee that the pedigree you receive is correct. It also has benefits in allowing you to join more than one sire to your herd, as the DNA will be able to determine the sire of each calf.
- Coat colour test (red factor and dominant black). A number of breeds have different colours accepted into their herd books. DNA typing for coat colour

Table 6.2: Heritability estimates for some traits in Angus cattle in temperate environments

Trait	Heritability (%)
Reproduction	
Conception	0–5
Days to calving	0–10
Gestation length	15–25
Maternal ability	20–40
Calving ease (heifers)	15–60
Semen quality	25–40
Scrotal circumference (18 months)	20–50
Serving capacity	15–60
Conformation and growth	
Body length	25–35
Birth weight	35–45
Weaning weight	20–30
Milk yield	20–25
Mature cow weight	50–70
Carcass (US)	
Intramuscular fat %	15
Eye muscle area	20–25
Dressing %	15
Retail meat yield	29
Yield % carcass weight	49
Other traits	
Eye cancer susceptibility	20–40
Temperament	25–50

Source: Adapted from G Fahey, D Boothby, G Fordyce, M Sullivan (2000) *Female Selection in Beef Cattle*. Queensland Department of Primary Industries, p. 22.

allows you to differentiate between homozygous and heterozygous black animals and to determine the incidence of the wild type allele. It assists in choosing the colour of your offspring but will not guarantee it, as red is a recessive gene with black being dominant. The red gene can be expressed as a true red gene or a wild allele (see Table 6.3). Other breeds such as Dexters also have a dun colour.

- Horns: an animal can be horned, polled or scurred (small horns not attached to the skull). The polled gene is dominant to the horned gene. The scurs gene is dominant in males and recessive in females. There are no available tests for scurs at the present time but there is for the poll gene.
- GeneStar testing: this provides molecular value predictions for core traits of feed efficiency, marbling and tenderness (see Table 6.4). The results from such tests are tools that can assist breeders in making genetic improvements. I have

found that my Auslines score very favourably in the feed efficiency tests: this reinforces the fact that the smaller breeds are able to produce more beef per hectare and require less feed to do so. I am also a fan of the tenderness gene. I have selectively used high-scoring bulls over high-scoring females and am convinced the result is evident in the meat of such offspring. The results from GeneStar testing are breed-specific, which means your animal's results are benchmarked across all the other animals tested, in your respective breed.

- Other DNA tests available include:
  - 94L myostatin muscling mutation for Limousin;
  - dilution factor (coat colour variant) for Hereford;
  - SureTrack: allows a producer to accurately trace a carcass, primal cut or meat product.
- Inherited disorders that can be tested:
  - pulmonary hypoplasia with anasarca in Dexter cattle;
  - chondrodysplasia (bulldog dwarfism) in Dexter cattle;
  - hypotrichosis in Hereford cattle;
  - idiopathic epilepsy in Hereford cattle;
  - arthrogryposis multiplex (AM) in Angus cattle;
  - neuropathic hydrocephalus (NH) in Angus cattle;
  - contractural arachnodactyly (fawn calf syndrome) in Angus cattle.

When considering these tests and the associated costs, you should be aware of the reliability percentages and remember they are just another tool to assist in your breeding program.

Q. We are finally getting into some beef sales, and therefore setting up some marketing. Part of this is looking into GeneStar testing, can you advise who you recommend we use for this? Any other tips you could share? Do you have all your cattle tested? Is there any additional worthwhile markers you would recommend?

A. Zoetis is the company responsible for GeneStar testing. It's a very simple process – just send in a hair sample from each animal you would like tested. I certainly don't test all my cattle, I mainly test my bulls. I'm not a fan of the marbling gene as the efficiency rate is only 26% and feeding grain for 70 days or longer will marble your beef anyway. I do think the tenderness gene is a very worthwhile marker and I note with interest that the small breeds seem to score quite highly in the feed efficiency marker.

Q. I am about to purchase our breeding stock (most likely Square Meaters or Galloways). One concern I have is to validate that the breeding lines of the animals are in fact what they claim to be. Do most associations

require DNA tests to register an animal with the Association? If not, how do we validate that the sire and dam for an animal are actually correct?

A. If the Association does not require parent verification DNA testing then you can get them verified yourself. You will need a hair tail sample of both parents and the animal you wish to purchase. A reputable breeder should have no issues with you requesting such tests.

Table 6.3: Possible genotypes and corresponding phenotypic expression of colour

Colour coat phenotype	Possible genotypes
Black	ED ED, ED e, ED E+
Red	ee, e E+
Various colours	E+ E+

There is a colour predictability chart on my website which you might find interesting if you are considering trying to breed a particular colour: http://vitulus.com.au/redline-2/examples-of-animals.

Table 6.4: Definitions of traits tested using Zoetis GeneStar testing

Trait	Definition	Reliability (%)
Feed efficiency	The difference between an animal's actual and expected feed intake, based on its bodyweight and growth rate. This helps you identify which animals need less feed to gain the same amount of weight.	30
Marbling	A visual assessment of the degree of intramuscular fatness in the rib eye muscle	26
Tenderness	The force required to shear a cooked steak after 14 days of postmortem ageing	49

7

Infrastructure required

Some basic infrastructure will be required if you intend to run livestock. The quality and cost will depend on the number of head, production goals, size of your cattle and the docility of your animals.

Fencing

Good fencing, from a management perspective, is important. Most behaviour in cattle is learnt, so good fences are vital. If cattle come from a property with poor fencing and have learnt that they can push through the loose strands of wire, then they will continue to do that for as long as they can. They will also teach their progeny to do the same. If, on the other hand, they have been educated from day one to respect fences then they should have no reason to even try to get out. Make sure your wires are tight, not slack – slackness allows stock to walk through easily.

If you are going to run a small herd with females, bulls and weaners then it's vital to ensure that the fences are able to contain your stock. Bulls chasing cycling females will do their very best to destroy a fence, and weaners trying to get back to mum can also be pretty persistent.

We have found that the most successful type of wire for our cattle has been hinge joint or ring lock (a special type of stock netting) with one or two strands of barb on top. The hinge joint keeps the small calves in and the barb wire stops the bulls trying to push over the top. Fighting bulls can be pretty damaging on hinge joint and in many instances, where we have bulls on either side of a fence, we have gradually replaced the damaged hinge joint with barb wire.

Five and six strands of barb wire work very effectively but in my experience three strands of barb are not sufficient for small breeds. Plain wire is of no use whatsoever – cattle will just push through it and walk wherever they want.

Electric fences work well but require regular checking and testing. They can be a much cheaper option on a semi-permanent basis and can be easier and faster to put up. The nice thing about electric fencing is the flexibility it allows. You can easily graze areas with just a couple of strands of wire, such as around the hay shed or your front lawn. Electric fencing works perfectly for strip grazing and we use it regularly in summer to feed off our forage sorghum. We allow the cattle to eat one strip at a time, moving across the paddock. This ensures no excess feed is trampled, so there is no wastage, and the previously eaten section is allowed to grow again.

It is very important to educate cattle to an electric fence system so they will soon gain a very healthy respect for all types of wire. Be careful when you first put cattle into an area fenced by electric wires. I have one paddock of 40 acres that is divided into six permanent strips by an electric fence system. When I first opened the gate and let my commercial cows in there they all took off at such a rate, kicking their heels up with excitement, that it ended up being a total disaster. They ran straight through the fences and I had busted wires everywhere. The secret to education is to introduce them slowly – I should have spent more time teaching them to respect an electric fence system.

Laneway systems work best and are also the easiest, allowing one person to move stock on their own. Our system is designed so that all the laneways feed into the yards. This reduces the time taken to muster the cattle by at least half and reduces the stress levels of both cattle and handler. It also means I can move my cattle all by myself.

When building new fences it is always a good idea to site gates in corners rather than the middle of a fence line. Corners prevent animals from running past the gateway and act as a forcing yard if you have large mobs.

There is a fabulous product called Clipex. With the Clipex fencing system there's no more tying, no more threading the wires through the holes – you simply push the wire into the grooves in the steel posts and the clip swings back to hold the wire in place. If you need to get a piece of equipment through a fence, you can just unhook the wires. It's so much faster to put up a fence using the Clipex system.

Tip: When setting up a portable electric fence, especially if you are connecting the cables to a battery, make sure you put the cables out of reach as calves love to chew them!

Q. Are the small cattle rough on fences?

A. No, and in fact they are much easier to contain than larger breeds. The biggest problem arises when people think that small cattle can be constrained by fencing that is inadequate to contain anything. Small cattle won't need such robust fencing as larger breeds, but they still require fencing that is in reasonable order. A client once rang me, concerned that she would have to sell the two exceptionally quiet 1st cross females she had bought from me. The cattle kept escaping into a neighbour's paddocks and the client assured me that her husband had fixed all the fences. I went to inspect, and discovered that their block was very high up on the hill with not very good quality pasture and the neighbour was down on the lush river flat. Their fencing was very very old – if I as much as put a finger on any of the star pickets it would practically lie parallel with the ground. Her husband had merely tried to straighten up the star pickets but the tension in the wire was so loose the cattle were easily able to push through it, especially when motivated by the much greener grass on the other side.

Tip: If your metal or mesh gates get bent for any reason, which happens over time, then take them off their hinges and lie them flat on a bitumen road or concrete floor and drive over them to straighten them out. It only takes five minutes and works exceptionally well.

Q. Do the calves tend to 'escape' a lot? I noticed that your fencing set-up was different from mine.

A. Calves tend not to venture too far from mum. If they do escape, they normally come back at feed time. I use hinge joint on all my boundaries to stop the calves getting out on the main road, as this is where they can easily frighten and run into oncoming traffic. Our riverbanks are fenced off to stop the stock causing any erosion and to allow the banks to rejuvenate with natural species. We used five strands of barb wire in preference to hinge joint so that in times of flooding the fences will be less damaged by floating debris. On one occasion a newborn calf had fallen asleep beside the fence and slipped under the bottom wire. She was having problems getting back to mum but was easily found due to the bellowing of both mum and bub late in the afternoon at feed time.

Paddocks

The decision on how many paddocks to have, and their size, will depend on what type of enterprise you plan to run. Are you are going to breed your own animals and how will you manage bulls and weaners? You will need to have enough paddocks so that you can segregate heifers from bulls from the time they are weaned until joining age. You may also need to keep those weaners away from their mothers for some months until they are fully weaned.

Permanent electric fencing set up for rotational grazing.

Hinge joint fencing with two strands of barb wire on top.

These paddocks are designed to join dirt laneways which then feed uphill to the yards.

Each paddock will need to have adequate protection from both heat and cold either naturally in the form of trees or man-made such as shelter sheds. Animals will need access to a good supply of clean water. In very large paddocks it may be necessary to install two watering points to allow cattle to graze all available pasture effectively, as cattle do not like to move too far away (5 km but some producers prefer 2 km) from their water supply.

Yards

The size and breed of your cattle are two of the factors that should be considered when designing cattle yards. Small cattle don't require expensive and extensive yard systems. They certainly don't need heavy-duty high rails or crushes. Your requirement will also depend on how quiet your cattle are. If you can call your animals with a bucket of feed, catch them with a halter and tie them up with ease, then nothing much more than a yard is required. In this case, light framed portable yard panels will work perfectly.

If your cattle are not halter trained then you will need some basic equipment as there are times when you need to restrain an animal, for example to administer vaccinations or seek veterinary assistance. You need to consider what functions you wish to perform in your yards, as this will affect your design and layout. It's preferable to have a race which feeds into a crush or head bail. If you don't want to go to the expense of a crush, then a walk-through head bail will suffice. At least it will allow you to restrain the animals without putting yourself at risk of being injured.

If you have a very small number of cattle then yards can be built out of gates, very effectively.

Table 7.1: Various yard functions

Function	Facilities/options
Drafting	Drafting race or yard
Vaccinations, ear tagging, tattooing, treating eyes	Head bail
Ear tagging, tattooing, treating eyes	Scoop
Using pour-ons or drenching	Race and/or head bail
DNA hair sampling	Race or crush
Weighing	Scales
Loading and unloading	Ramp
Dehorning	Head bail or calf cradle
Castration	Head bail or calf cradle
Veterinary: pregnancy testing, AI	Crush with veterinary access

The minimum recommendation is one yard, a race and head bail. You will also need a loading ramp if you intend to move cattle onto and from your property. There are some very good portable ones with adjustable ramp heights which can be used for a range of transport options from body trucks to small trailers.

You will find that over time your cattle will learn your yard system. They will learn how the yards flow and what is expected of them. Some of my very quiet cows walk into the crush and stand with their head through the head bail waiting for it to be closed!

Functions

Most yards will need facilities for receiving and holding stock, drafting, loading and unloading, weighing and basic herd health. Pregnancy testing, artificial breeding, dehorning, branding, castrating, vaccinating, weighing and spraying may also be considered. Table 7.1 shows various yard functions with the facilities required.

Hazards

The hazards involved with handling cattle include injury due to kicking or charging, crushing against handling facilities and illness transmitted by cattle (zoonoses). To reduce the risk of such accidents you will need to consider the size, breed and temperament of the animals being handled, the competency of the handler and the design of the facilities. You need to design the yards and consider the way cattle behave, drawing them through the yards and reducing distractions. You also need to design them for the safety and ease of handlers, using self-latching gates, easy access and surfaces that reduce the risk of falls and trips.

Site

If you do have to build a set of yards, consider making them centrally located in your property so that cattle can be mustered easily. Being accessible by an all-weather road is important. Other important factors include being built on a slight slope, the direction of the slope and the way the yards face in relation to the sun. A slight gravelly slope is preferable to the bottom of a hill on black soil, so that the yards can drain and dry out faster after periods of rain. Consider the time of day that cattle will be worked in the yards, as light and wind factors can affect the way stock behave. Natural animal behaviour should be taken into account; for example, cattle prefer to run uphill and towards the light. They tend to baulk at dark shadows.

Yards will need to have all-weather access, especially to the loading ramps, which is why they are often located close to the property boundary. If the yards are located a long way from the home, it's a good idea to have lockable gates at least at the end of the loading ramp, to deter stock theft. Natural shade is an advantage, especially if stock have to be left in the yards for long periods in hot conditions.

Environmental considerations

Covered work areas, especially the race and the crush, make the stock and handler more comfortable and therefore less stressed. They also provide the handler protection from the sun. Concrete floors and covered crushes reduce the dust and subsequent risk of infection when cattle are being vaccinated or castrated or veterinary treatment is being administered.

Shade

It is good idea to make sure your yards have some type of shade to prevent against heat stress and skin damage. Shade can be provided in the form of trees, shadecloth or covered sheds. Trees need to be hardy to withstand high levels of urea and acid soils. In colder climates, deciduous trees maybe preferable. These will enable the sun to penetrate during the cooler months to provide warmth, and assist in drying out the surface in wet times. Possible species include the plane tree, white cedar, *Schinus molle* (pepper tree) and *Ficus benjamina*. Remember to site trees so they don't interfere with the flow of the yards, fences or other structures as they grow. Shadecloth can be erected easily over yards but check the level of protection offered as it can vary from 15% to 100%. Permanent structures are the most effective but also the most expensive; they are a long-term investment. A high roofline is preferable to allow the air to flow through.

Size

The yards need to be adequate for the biggest herd and the largest animal which will be handled. For large-size breeds the recommended space is 1.5–2.2 m^2 per adult beast. Forcing yards should not be too large (a maximum of 6.5 m), otherwise two or three people will be needed to force the cattle up. A top rail of 1.6 m is the normal recommended height and more than adequate for small breeds. The recommended race length is 1.6 m per adult beast with the minimum length for up to 10 head being 3 m. Recommended width is 660–710 mm for wide European breeds.

One problem for me is that my race needs to be wide enough to accommodate my large commercial cattle, which means it's wide enough to allow my Auslines to turn around inside it, which can be very annoying. No doubt in years to come someone will manufacture adjustable race panels! Be warned: don't build a permanent race and yard system to accommodate small breeds if later on you might consider having other breeds as well – they just won't fit.

Design

Well-designed yards can make cattle handling very easy for one person. Poorly designed yards can make it a nightmare as well as being extremely dangerous. The real test of a good cattle yard design is to see if the cattle walk straight into the race without baulking.

- Cattle will move more easily in a curved race than in a straight one, as they think they are going back to where they came from.
- Solid sides work better, as the cattle can't see any distractions outside.
- It is a good idea to make sure your yards have shade and water so animals can be left there if they need to be treated on a regular basis.
- Drafting yards of hexagonal or octagonal design, which are centrally located, enable the removal of calves from cows, selecting cattle for sale, marking or branding and separating different mobs.

When designing yards remember to keep safety in mind.

- It is better to be able to handle cattle from outside the race.
- Try and avoid having to handle cattle through gaps in yard rails as this increases the risk of crushed hands and arms. Reversing cattle will not see you and/or not be able to stop.
- Don't make the race so narrow or wide that the handler has to intervene.
- Head restraints in the crush will reduce the risk of accidents when tattooing or ear tagging.
- Avoid acute angles in the yards where cattle can get trapped.
- Inward-swinging gates are best, except for the entry gate.

The advantage of the basic plan is that extra yards can be added as required, to accommodate a larger herd.

Equipment

The equipment you buy will be determined not only by price but also by the functions you will be performing, such as castrating calves, dehorning, pregnancy

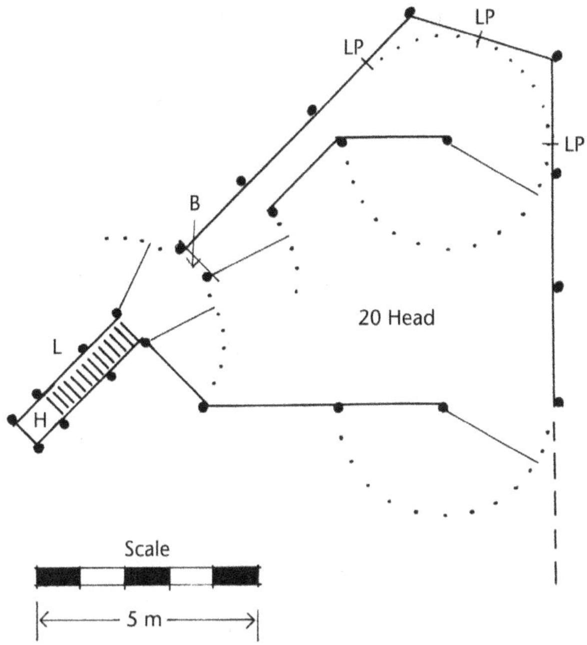

These yards are designed to handle a maximum of 20 head. Drafting capacity is enhanced by yards on both sides of the race. B = head bail, L = loading ramp, LP = locking point. Source: Adapted from *Farm Cattle Handling*, SCARM Report No. 35.

testing, weighing etc. The best way to compare products and prices is to go to an agricultural field day. Manufacturers often offer special field day prices as well.

Crush

The most important piece of equipment in your yards will be a crush. Once again, function will determine what features your crush must have. Functions such as pregnancy testing, AI, embryo transplant, vaccinating, tattooing, dehorning and castrating will determine if you need vet sections, baulk gates, a squeeze mechanism and split side gates. Prices can vary from $3000 for what is often called a hobby farmer model up to $20 000 for a crush which is hydraulically operated and remotely controlled.

A veterinary section is recommended and will be required for AI and embryo transplants, as the cow will need to be restrained in a clean, shaded and stress-free environment. The head bail is required for AI, so the cow needs to be contained in the crush with the AI technician having access from the rear end. The benefit of a vet section is that it allows access to the animal in the crush while also protecting the handler.

We have a parallel squeeze on our crush (one that squeezes equally from both sides) which not only prevents the smaller animals from turning around in the

The Hobbymaster crush is a basic crush with no vet section but I like that the bottom panel is sheeted, as this prevents lower legs getting out. The undercover crush is ours: it has a separate vet section, access panels on both sides with split gates top and bottom, a parallel squeeze, head scoop and scales.

crush but restricts the amount of movement available to them. This makes them feel more secure if we have to do veterinary treatment. Some crushes offer what is called a bayonet side squeeze but I am not a fan of these.

Split side gates are very useful for access while vaccinating animals. Putting hands through crushes is a recipe for injury especially if the animal moves in an upward direction, jamming your hand or arms against the bars.

Baulk gates sit in front of the crush and prevent an animal getting away. They are used when you have particularly difficult cattle to catch and are great when you miss an animal in the head bail! We don't use baulk gates due to the design of our system.

Scales

The second most important piece of equipment is electronic scales. You will need two parts (i.e. load bars and an indicator); these are sold separately. The price will depend on whether you have lightweight or heavy-duty weigh bars and what functions you want your indicator to perform. Do you just want them to accurately weigh each animal or do you want them to create files for each session, give you weight gains for each animal, calculate and view carcass weight, track an animal's performance through its NLIS tag, transfer data to your PC? The list is endless. Prices vary, from the cheapest set at approximately $1500 to the full works in excess of $10 000.

Scales also allow you to perform a number of useful additional functions such as monitoring animal health, determining weaning weight, selecting animals for sale and slaughter and evaluating breeding performance for genetic selection. They can assist you to tailor your feed and treatments as required and monitor the weight gain of offspring to help make better genetic choices. It's also very nice to know the weight of your animals prior to selling them to the saleyard or abattoir, so you can make sure you are getting paid for every kilogram of meat.

One function for which we use our scales every day is to accurately measure drenches, pour-ons and medications. As nearly all are based on an application rate per kg, accurately weighing your animals prior to medicating them means you avoid wastage, reduce costs and maximise the health of your animals. Most of these products are very expensive and until you are experienced with cattle weights, if you don't use scales you could be literally wasting hundreds of dollars. Likewise, if you undertreat then applying these products is pointless.

We recently updated our scales to a TruTest set. The two functions I use most are the ability to retrieve the last weight of that particular animal and its daily weight gain since last weighed. I feed this data into my software program.

Ancillary equipment

The list of available extras is constrained only by monetary considerations. All work extremely well and can help reduce the manpower, stress and costs involved in handling livestock.

- Head scoop: the photo of our crush shows that a swing arm called a Super Scoop is attached to the front. This head scoop is designed to restrict side-to-side and downward movement of the animal's head. The rubber-lined scoop holds the animal's head in position, which makes it ideal for ear tagging, tattooing, eye treatment etc. It also provides a safer working environment for both the animal and the operator.
- Round forcing yard: we use one of these systems purely for safety reasons. It saves anyone having to get in behind the cattle when pushing them into the race.
- Automated drafting systems: normally installed only when large herd numbers are involved.
- Calf cradles: these are marketed as safer, easier and faster for handling large number of calves for marking, branding and dehorning.
- Integrated NLIS ID equipment: this can range from readers, scanners and wands to fully automated race readers. Electronic devices can allow producers to keep track of individual weights, health and joining records of animals. Most indicators on scales offer this facility.

Loading facilities

If you move livestock onto and off your property you will need a loading ramp. There are a couple of options to consider. My recommendation is a galvanised adjustable ramp, which will allow you to load from a box trailer through to a truck.

- The slope to the ramp should not exceed 20° and the ramp should be approximately 900 mm wide. While this may seem wide for small cattle, there may be times you will need to load and unload regular sized cattle so the ramp must be able to accommodate these as well.

- The ramp should lead to a level entrance so that cattle can balance before entering and leaving the truck.
- Ensure that the ramp has non-slip footing which does not create undue noise.
- Make sure the base is closed in to prevent legs getting outside the ramp.
- Make sure there are no large gaps between the back of the truck and the ramp, in which an animal's foot can get caught and easily break in its haste to get off a truck. You will need to make sure that the truck is able to back straight onto the ramp, so the posts should be even and straight.
- Cattle should not go up a ramp into direct sunlight.
- Some brands of ramp include the option of a side walkway, which assists in loading cattle.

We have two ramps side by side. They both feed into the same race. The portable metal one on the left is adjustable for horse floats and trailers, while large livestock carriers use the timber ramp.

- Some brands include the option of wheels, making them truly portable. If for some reason the truck cannot get to your yards, with an ATV or tractor you can take the ramp to the truck.

Water

The amount of water consumed by a cow varies dramatically and depends on a number of factors including temperature, humidity, availability of shade, lactation, moisture content of the food and growth. Cows drink on average about 1 L for every 10 kg of bodyweight plus 15 L for every 5 L of milk produced. This averages out to 40–70 L of water per day for a small-breed cow, but of course will vary due to the conditions listed above. It is known that a 10° increase in temperature (e.g. from 25° to 35°) can almost double daily consumption.

If you observe cattle in the paddock they normally drink twice a day, usually between 10am and 11am then late in the afternoon. Once again this can vary with temperature, humidity and how far cattle have to walk to get to an access point.

Clean water is essential. It has been shown that cattle that have access to clean palatable water gain weight faster than those that have to drink dirty water. Dams that are low in water supply are often the worst as the animals urinate and defecate all around the edges every time they drink. As the dam's water supply is reduced, it becomes more and more contaminated. Reduced water intake can result in lower forage consumption and, consequently, reduced weight gains. Be careful when using troughs, as anaerobic bacteria can build up on the inside of pipes.

Field trials in the US have shown that much of the water drunk from dams can reduce livestock performance. One trial showed that heifers with access to fresh water delivered to a trough gained 23% more weight than heifers drinking from a dam.[1] It also found that cattle that had a clean water supply spent more time grazing and less time resting. In another study over a 33-day period, cows drinking from a dam lost 0.2 kg and their calves gained 26.3 kg while cows drinking fresh water from a trough gained 7.4 kg and their calves gained 33.9 kg. A more recent study found that 92% of the time cattle prefer to drink from a trough rather than from a stream or dam.

In addition to checking the cleanliness, a water analysis is recommended before water is used for stock, domestic or irrigation purposes. This will check the salinity of your water as well as specific ions such as calcium, fluoride, magnesium, sulphate, calcium and nitrates, which can cause health problems. These problems include loss of condition, scouring, teeth decay and gastrointestinal problems. For example, excess fluoride can occur in bore water and can be a problem particularly for young stock, as well as for human health. Samples can be sent to a private

We use round troughs that hold their shape better. Note that both have their floats protected as well. We have installed a smaller trough as we found that newborn calves were unable to reach into the large troughs. Every couple of years we add new gravel around the bases, not only to prevent erosion but to maintain the even levels.

company, Landcare and Catchment Care groups. Even try your local authorities, as some offer a water testing service.

At least 60 cm of accessible linear water space is needed per 10 head of cattle. When installing troughs make sure they have a bung so they can be cleaned out regularly. We do ours on a monthly basis with a good scrubbing brush to remove the build-up of algae. Trough block, available from your local produce store, is added each time to help reduce algae and slime build-up. Make sure you read the application rate and don't use it if copper toxicity is known to occur. Trough block will last six to eight weeks. Goldfish are a good natural way to help keep algae under control. Polythene troughs are more heat absorbent, which promotes algae growth.

Round poly troughs seem to hold their shape better than the rectangular ones. Concrete troughs keep water the coolest. Concrete troughs also have an

Tip: You may need to consider height when purchasing your troughs, allowing for small calves to drink from them. Make sure you buy troughs where the float is protected and covered, otherwise cattle will play with and eventually break or damage it while pushing and shoving for a drink.

This is an example of a rectangular poly trough on our property that has lost its shape over five years. The problem is that the water then starts to run over the sides, which means that the float has to be lowered. The timber rails above prevent animals trying to walk over the top, and allow one trough to service two paddocks.

advantage in that if they run dry the cattle have great difficulty moving them, unlike poly troughs which can be seriously damaged by stock if their water supply is restricted.

If possible, in warmer areas it is a good idea to position troughs under some natural shade. This way in summer the water in the troughs won't get too hot, allowing the cattle to drink clean cool water.

Shelter

Under government regulations, livestock owners have a duty of care to provide adequate shelter for their cattle, especially shade during the summer months. In extremely hot conditions, grazing animals have to expend a large amount of energy on cooling themselves rather than converting food into protein. Cattle understand this need, which is why they graze in the early and late parts of the day and rest during the heat of the day. If animals suffer from heat stress for long periods then damage to muscles and organs can occur, resulting in death. Signs of heat stress include excessive panting and the cessation of ruminating.

An old shed like this makes a perfect shade shelter for the stock. It's nice and high with no sides, thus allowing the air to flow through.

My father told me that a contented herd would be resting in the shade of the trees by 11am, meaning they had enough feed in their paddock to allow them to rest in the hotter part of the day.

Trees offer the most effective and coolest shelter, but reducing pressure on trees is also important. Single trees are most in danger due to concentration of urine and manure around the base. Trees will need to be fenced off for at least five years. Otherwise the cattle, bulls in particular, will eat them (not because the trees are tasty but just because they can), rub and scratch up against them, often ringbarking or breaking them altogether. A good idea with single trees in a paddock is to trim the lower branches, which will force the cattle to lie further away from the base as they chase the shade. They will also have to move more regularly to obtain shade, due to the angle of the sun changing throughout the day. This reduces the manure overload at the base of the tree, which can be severely detrimental and sometimes cause the tree to die.

Shade sheds can be built until such time as timber belts are established. The roof should be high to allow the best ventilation of the cattle's body heat. There should be no sides, so that breezes can flow thorough. In hot dry areas align the structures north–south, but in hot humid areas an east–west alignment is better.

An animal's energy requirements may more than double in cold weather, just to maintain the animal's body heat. The smaller the animal, the greater the increase in requirement. Shelter increases both temperature and humidity, with the largest increase at a distance of about five windbreak heights from the windbreak, but this extends to a distance of about 12 windbreak heights downwind.

Irrigation

Dams

If cattle have access to a dam you will quite often see them standing in the water on a very hot day.

Sprinklers

Some farmers, particularly those with dairy herds, use irrigation sprinklers which their cattle can stand under on extremely hot days. You would need to move these on a regular basis to prevent boggy patches and possible associated foot problems.

We use the K-Line system, which is perfect for a small acreage. It is a system of small plastic pods protecting a small sprinkler attached to special K-Line low-density poly pipe. It has a number of advantages over traditional systems, being made from durable polypropylene.

- It is very cheap when compared to other irrigation systems.
- It is virtually indestructible – you can drive over it and cattle can stand on it with no damage being done.

K-Line system working on our property.

- It is inexpensive and easy to maintain. We have had a system in use for over 10 years and there have never been any serious problems. The only parts that we have had to replace are the sprinkler heads that cost less than $10 each.
- It is exceptionally easy to move. You just need a four-wheel motorbike or vehicle with a towbar. There is no need to purchase a tractor. One person can move it by themselves within 10 minutes.
- It can go anywhere such as through gullies, around trees and places other irrigation systems cannot be used.
- It will run on low-pressure pumps so that the amount of line used can be adjusted to suit the available level of pressure. It is designed to distribute water via a slow absorption method.

Q. My family lives on 4 hectares and I would like to have some cattle. I was unsure as to what breed. At present, I'm in the process of developing the pasture and setting up paddocks, yards etc. I didn't want to rush in and buy the cattle and then put them onto substandard pasture and arrangements, even though they (the cattle) are extremely appealing. To date, I have four main paddocks fenced (with five-strand wire, three of which are electric). I have a bit more fencing to do before bringing the cattle home. The additional fencing will make it easier to move the stock from one area to another without too much stress for either them or me (yes, they'll be mainly my babies!). What other things should I be planning for in order to provide the best possible environment for the cattle?

Another concern I want to cover before bringing home my (babies – you're definitely right), is while we don't have a dog problem in our area that I'm aware of, we do have dingoes. As they're only so little, do you lock up your young ones in either a yard or paddock close to the house? That probably sounds like a dumb question but I thought I'd show my ignorance and ask it anyway.

A. I have never had a loss to date but we do have dingoes. I don't lock up the young ones but I do bring them in closer to the home paddocks when the cows are calving so I can keep an eye on them and check them daily. Cattle are very protective and will form a circle to protect the young ones. They will also chase wild unfamiliar dogs. Fox lights (see p. 207) are a good deterrent for dogs as well.

8

Basic cattle husbandry

Initially you will probably require a vet to assist you with everything, but as you become more familiar with the processes you will learn to do the majority of basic husbandry, such as vaccinations, yourself. Find a vet who enjoys and is experienced at working with cattle, is willing to teach you and work closely with you. They will be more than willing to assist, as they really don't want to come out to give a simple vaccination.

This chapter will explain some of the basic things you will need to know, and what you can do yourself very easily. Other, more complicated practices you can learn over time. At first it will be daunting, but it really is very simple and just requires some practice to give you confidence.

When your first calf is due to be born you will probably be a nervous wreck, checking every five minutes, making sure that everything is okay. When the calf is born you will be over the moon with joy, posting on Facebook and Instagram and emailing your family and friends about the new addition.

One client left his first commercial purchase at my stud to calve. I was given strict instructions to inform him immediately of the birth and all details pertaining to the calf's arrival. On the day it was born I rang the number he had left; the girl who answered the phone informed me that he was not there. 'Any messages?' she asked. 'Just tell him that his baby was born today. It's a boy and it's black', I replied. 'Oh, okay' was the response. I'd spoken to my client's daughter-in-law, who knew nothing about his cattle purchase. He rang me that night with great amusement: his daughter-in-law had immediately called him at work to whisper very quietly about a girl who had rung about the birth of his black baby boy!

Health and safety

Workplace health and safety should be high on your list of priorities. The agricultural sector is rated as the second most dangerous occupation in Australia, with mining and transport being the first. Cattle producers need to be aware of their obligations and the associated regulations in each state.

Recent statistics from the Australian Centre for Agricultural Health and Safety show that there are on average nearly three work-related fatalities every month in Australia in the agricultural sector (Table 8.1). A study of farm-related deaths between 1989 and 1992 showed that 16% occurred on meat cattle properties.

Since 2001, over 190 Australians have died in quad-related incidences. Quads are now the leading cause of non-intentional injury death on Australian farms

Table 8.1: Agents of injury causing death as reported in Australian print media, January 2013–June 2014

		January 2013 to June 2014		
Category	**Agent**	**<15 years**	**>15 years**	**Total**
Animal	Cattle		1	1
	Horses		1	1
Electricity	Powerlines		2	2
Farm structure	Dam	2		2
	Roof		2	2
	Windmill		1	1
Farm vehicle	Aircraft		1	1
	Farm vehicle NEC*		1	1
	Helicopter		1	1
	Motorcycle	1		1
	Quad	2	10	12
	Side by side vehicle			
	Trailer		1	1
	Truck		1	1
	Utility	1	3	4
Materials	Material other NEC*		1	1
Mobile farm machinery/ plant	Tractor		9	9
	Earth-moving equipment	1	4	5
	Not specified		3	3
Other	Firearm		1	1
Working environment	Trees being felled		1	1
Total		7	44	51

*NEC = not elsewhere classified.

Source: Adapted from *Australian Farm Deaths and Injuries Media Monitors Snapshot January 1st to June 30th 2014*. Australian Centre for Agricultural Health and Safety.

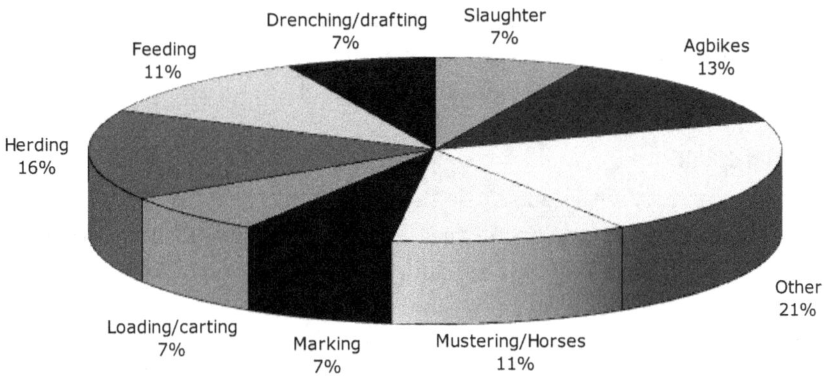

Cattle activities resulting in injury. Source: Australian Centre for Agricultural Health and Safety.

(outranking tractors). Deaths are evenly distributed between rollovers, where asphyxiation/crush injury are common, and non-rollovers, where the victim is flung onto a hard surface as a result of a quad bike crash. Almost nine out of every 10 rollover deaths occur on a farm. If you are an employer or in control of the farm workplace, you have responsibility under work health and safety laws to provide safe systems of work for workers and visitors to the workplace, including the operation of quads. Make sure everyone wears a helmet. My advice is to buy a UTV, as they seem to be a safer option and certainly can't go at the same speed or turn as quickly and tightly.

When assessing cattle activities resulting in injury, evidence seems to suggest that these accidents arise from a combination of lack of knowledge and understanding of cattle behaviour, stress due to inadequate facilities and poor work practices.

Cleanliness is essential to reduce the risk of infection in animals and the transmission of zoonotic diseases to the handler. All animals should be vaccinated for leptospirosis and all workers vaccinated for Q fever. Q fever vaccinations are available through the Health Department of each state and should be made available to all cattle handlers, particularly to workers not previously exposed to working with cattle.

Behaviour

An understanding of animal behaviour is very important, as the greater your knowledge the better your ability to predict an animal's response. While some handlers have a natural ability, for many the skills are gained from observation and practice. Over time you will learn to be instinctive and sense when cattle are ill, distressed or hungry, and know what the signs are. Remember that cattle are clever and can learn to be very cunning. One of the world's leading experts on cattle behaviour, Dr Temple Grandin, points out that fear memories in cattle can never be deleted, only reduced. We should never punish fearful behaviour; instead we should reinforce and reward the behaviour we want.

Human interaction

The more you interact with your cattle the quieter they will become, and the younger you start the better – as long as the interaction has positive memories for them. If every time you go out to the paddock you yell, shout and whack cattle with a stick they will very quickly learn to take off to the other side of the paddock to avoid you. If, on the other hand, you move gently, talk to them quietly and offer some reward, such as a biscuit of hay, each time you visit they will be eager to see you and come running towards you. They will learn to recognise your voice and even your vehicle. It is much easier to muster cattle by calling them rather than having to chase them into the yard.

You must be firm with your cattle,, as they are used to being dominated. If they headbutt you, smack them to let them know this is unacceptable behaviour. How many times have we seen a tiny calf at a show start to headbutt children after a few days of continual patting, much to the public's amusement? If this behaviour is left unchecked then that young calf will grow up thinking it's acceptable. I can assure you it won't be so funny when he is a fully grown 500–700 kg bull.

Herd hierarchy

Cattle are social animals that operate in large, highly organised groups and because their herd instinct is strong their social structure is established at a young age. Within a mob there is always an order of dominance, often seen in action at water or feed troughs where certain animals are always first to drink or eat. More timid animals will stand back until the others have finished. This order of dominance can cause stress, especially when the animals are confined in small spaces such as yards, feedlots and trucks. This is why there is always some fighting whenever you mix cattle into new groups: it allows them to work out the pecking order and establish their very defined social dominance order. The most dominant animals will test the new animals until they find their spot in the hierarchy.

Young bulls likewise soon learn that older bulls are very definitely the boss, but be careful with senior bulls. Often they will fight until one is seriously hurt, sometimes resulting in severe damage such as broken legs or shoulders. Don't put older bulls of similar age and weight together in small paddocks or yards. They will keep fighting until one has assumed dominance, and this is often to the detriment of the other bull. Bulls can be uncontrollable and highly aroused when fighting – and totally oblivious to anything else, so keep your distance. Don't jump in to try and separate them or you may end up with serious injury. Overcrowding often causes problems and it is recommended that large breed bulls in yards be allowed 6 m or more of personal space.

Dominance behaviour is not restricted to males. Females don't seem to take so long to work out the order and the more timid females learn to keep their distance. The behaviour is often most noticeable when a female is in oestrus and on heat. She is trying to prove that she is the most dominant to attract the male, as in a natural setting the dominant male breeds to the dominant female.

Know the signs

Learn to recognise and understand the signs that cattle make, as understanding their behaviour may help reduce injury and may even save your life. If they are cranky they will swish their tail from side to side, dig up dirt with their front feet, hold their head up and stare at you and sometimes snort if you get too close. Watch their ears as well – as cattle are very sensitive to sound these are a good sign. One golden rule is never stare cattle down. Back off and don't invade their personal space, as this will allow them to calm down.

Contented and healthy cattle will have a good stretch when they get up, then revert to their normal posture. Stressed cattle can have arched backs and lowered head and ears.

Protective mothers

Always be careful of cows and newborn calves, especially of cows that have calved for the first time, no matter how well you think you know them. They will protect their calves from all intruders so be very careful about picking up or patting a newborn. Assess the cow, talk to her quietly and watch her to see if she will let you get close to her baby. Move slowly and calmly and reassure her with your voice. She will give some very serious signs if she is going to try and stop you, by charging at you. Sometimes a charge is just bluff but be warned, sometimes it isn't! Watch her eyes the whole time. Remember she is working on instinct, don't confuse this with bad temperament. Try to keep dogs away from cows that are calving, as cows will often react very strongly to any type of dog nearing their young irrespective of where you are. Often they will tolerate you, but nothing or no one else. This type of behaviour is generally not a major issue with small breeds, due to the large amount of human interaction, but there are exceptions to every rule.

Hiding the babies

Most cows will hide their newborn for the first couple of days away from the remainder of the herd, until they are strong enough to interact with the other cattle. They will 'plant' them in tufts of long grass or under a shady tree, coming back to feed them a drink of milk when necessary. Some mothers will remain quite close by while others will leave their calves there all day, only returning in the late afternoon to give them a drink.

It would be impossible to count the number of hours that I have spent looking for newborn calves. I have walked acres of paddocks from one end to the other knowing the cow has hidden her calf somewhere but unable to find it. Cows have deliberately led me in the opposite direction, totally ignored me when I have imitated their calf call, or pretended they can't find their calf either. The best way to make sure she (and you) have found her newborn is to wait until evening as the mother will eventually track it down to give it a drink before the sun sets.

We have an old well on our property and, unbeknown to me, the roof on it had been dislodged in the 2013 flood. Some months later one of my commercial cows was bellowing continually and I assumed she had lost her three-month-old calf. I went over to see what the issue was. She led me to the well and stood there and bellowed at it ... then after five minutes she turned around and walked off. There was no visible problem and after searching all the paddocks I couldn't find the calf. Some four or five days later the calf floated up to the top of the well. The mother had been trying to tell me what had happened.

If you are concerned and you think a cow has lost her charge, check to see if she is restless and continually bellowing to her calf. Also check her udder to see how bagged up with milk she is. If she has not fed it for a while, her udder will be tight and perhaps even leaking milk. Sometimes a calf will get out under a fence, fall into a gully or become lost due to a sudden disturbance that has made it run in the wrong direction. A good indication that a calf has been lost is that the mother will become very distressed and usually won't give up bellowing until she has found it.

One exceptionally hot and humid February my calves were suffering from dehydration due to the fact that their mothers were leaving them alone all day without a drink. The solution was to find the calves at about lunch-time, get them on their feet so their mothers could see them then make the mothers return to give them a drink. To prevent this happening again I changed my autumn calving period to begin on 1 March, rather than on 1 February which had been the case in previous years.

Day care

Cattle, being herd animals, have devised an ingenious system of day care. One or two females are left in charge of the nursery for the day or a part thereof, allowing the other mums to go off and graze. Large herds of cattle often have two day care centres in separate parts of the paddock, depending on the herd's social structure. This day care centre will normally be under the shade of a large tree or some other protected place.

The babies are normally booked in the late morning and the mothers come and go, depending on the calf's age, to give them a drink. Each day the carers are rotated, ensuring that all mums get a break and more importantly are able to graze adequately. It works extremely effectively. Cows with newborns may often take a couple of days before they decide to use the day care facilities. As mentioned previously, they keep their newborns close by and away from the rest of the herd until they are strong on their feet.

The day care system. These two cows were baby-sitting eight calves under the shade of a large tree.

Moving and handling

Knowing and understanding the natural instincts and common behaviour of cattle is the foundation of low-stress stock-handling principles. It's not rocket science, it's just common sense and may require a change in attitude about how stock should be handled. Good stock handlers are calm and able to move quietly while maintaining control of their stock. They are also able to understand the social behaviour of cattle, which will vary depending on breed, species, age and sex.

Cattle like to be moved in mobs, they don't like to be singled out. Cattle also like to keep other cattle in sight and because they have a field view of 330° they can see from almost all directions. If you need to draft some cattle out from a paddock it is often much easier to bring the herd into the yards and draft from there, rather than trying to draft them out from the rest of the herd in the middle of the paddock. If you have one animal that continually runs out of the mob or is a nutcase then get rid of it, as it will teach the other cattle bad habits. Don't go wildly chasing animals that break away or escape – more often than not, if left alone, they will try to return to the mob or the yards.

Some years ago I attended an induction day for exhibitors to witness the weighing-in of cattle for one of the most significant hoof and hook competitions in the state. There were over 100 head and numerous handlers to weigh, tag and draft the various animals. There was no evidence of any low-stress stock-handling principles being employed and most of the cattle were in a high state of arousal. This was due to the fact they had all come from different properties and were being yarded together for the first time, were being worked in unfamiliar yards with far too much noise, and were being weighed, ear tagged and injected. The experience was most unpleasant. During the course of the day someone accidentally left a gate open and two of the steers escaped. The commotion, by so-called professional cattlemen, that followed this escape had to be seen to be believed. They were hollering, yelling, chasing, banging gates, using dogs and making so much noise that, not surprisingly, the two animals took off to the scrub country and have never been found. A financial loss I'm sure the exhibitors could have done without. If those so-called cowboys had quietly left the steers on the outside of the yards to calm down they would have easily been able to bring them back in without any problems.

Learn to understand the flight zone

The flight zone is the distance an animal will allow a person to approach before moving away, in other words it's their personal space. It is essential to be aware of the animal's imaginary safety zone, and of its flight zone. If their space is invaded cattle will try to escape, and they also like to turn and face that pressure if it becomes too much. The flight zone is bigger when an animal is excited or stressed. Minimising stress when moving stock is all about applying the right amount of pressure and then releasing it.

An animal's flight zone is determined by three factors: genetics (temperament), amount of human contact and the quality of that human contact. Animals which are handled roughly or irregularly will have a much larger flight zone than those that are handled gently and regularly. Usually the flight distance is about two to three body lengths, but this will vary depending on an animal's past experience with humans. For wild or feral cattle it maybe 200–300 m, for feedlot cattle it may be 1–5 m and very tame cattle, such as your show team, will no longer have a flight zone. Sometimes very tame animals that no longer have a flight zone can be difficult to work with in yards as they have no fear and will try to do exactly as they please.

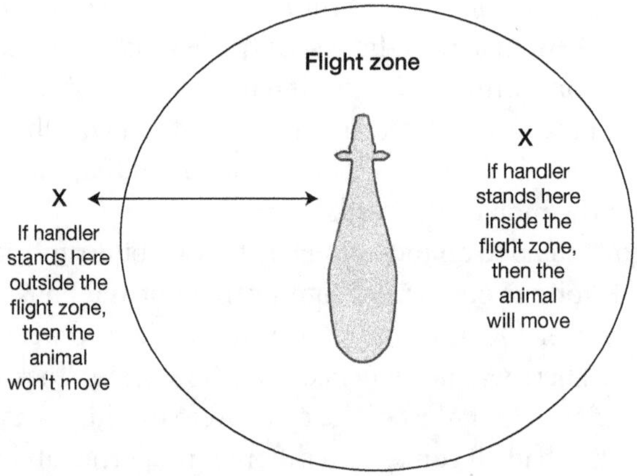

The flight zone. Source: Adapted from Temple Grandin, http://www.grandin.com.

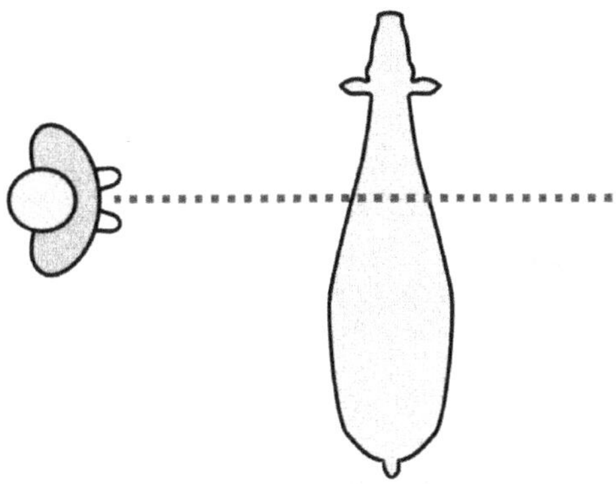

Shoulder point of balance. Stand on line and beast does not move.

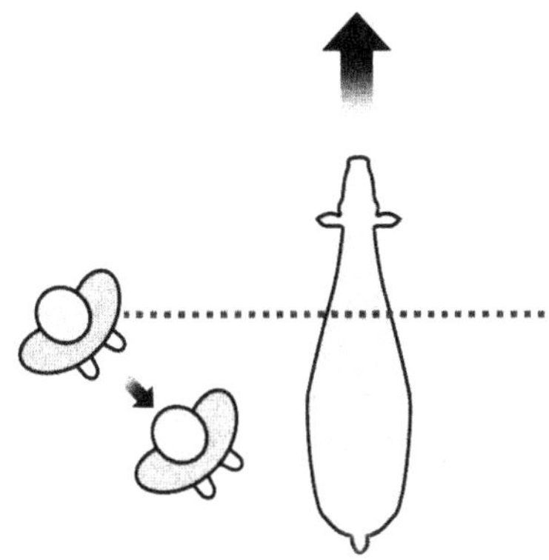

Moving behind shoulder point of balance. Move behind shoulder line and beast will move forward.

Learn to understand an animal's point of balance

Cattle have what is called 'a point of balance' which is about level with their shoulder. This is where the animal will move neither forwards nor backwards if approached. If you move in front of their shoulder they will go backwards and if you move behind their shoulder they will go forwards. This applies to all cattle whether when you are working in the yards, in the race or in the paddock. You will see it works very effectively in the race when you want the cattle to move forwards. Just walking slowly and directly towards the cattle, about 1 m away from the rails and in a straight line, will encourage them to move forwards as you pass their point of balance.

Cattle move more effectively if they can see the handler at all times, which is why it is important to avoid standing directly behind cattle as they will naturally turn to see you. The most effective point is to have the handler situated at a 45–60° angle from a line perpendicular to the animal's shoulder or point of balance. The same principles apply to moving small groups or large mobs of cattle. They will

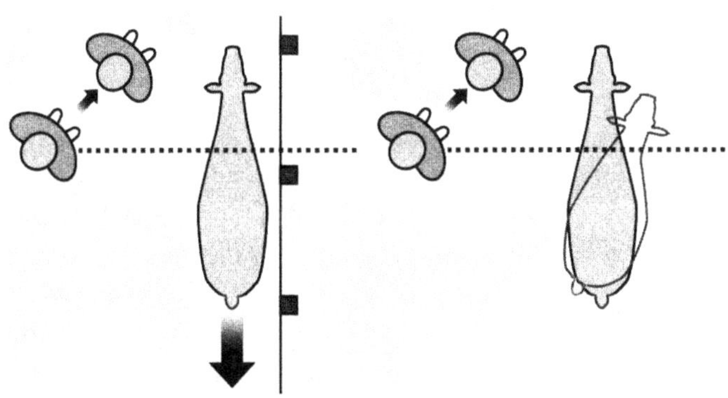

Moving in front of shoulder point of balance. If this occurs on the fence the beast will move backwards. With no fence, beast will move away.

have a collective flight zone and when the handler penetrates that zone they will move forward. By entering and then releasing the pressure the cattle will continue to move forward.

Yard techniques

Once you have brought the animals into the yards give them some time to calm down, especially if they are in a high state of arousal. This should be approximately 20–30 minutes.

Consider the features of the animals you are working with, such as the breed, size and temperament, whether they have horns, are males or females, are cows and calves or cycling heifers and bulls. Horned cattle are considered more dangerous than polled cattle and bulls are usually more aggressive than females, especially if there are cycling females in the yards with them. This is because their mind is on other things than moving through the yards and doing as you wish. Cows with calves are more protective in the yards as their young get lost and mixed up, and they and the calves can get quite distressed.

Cattle should move easily through the yards and enter the single-file race without hesitating. If the cattle baulk and refuse to move, then try to work out what it is that they don't like. Perhaps the entrance to the race is too dark, the sun is blinding them so they can't see, there are dogs in front of them, too many shadows, a change in the type of flooring, a hat or coat on a fence, some chain or a piece of string hanging down. Little things that you may not have even thought of, but if you get into the race and get a cow's-eye view you will probably find the problem. Cattle don't like the dark or shadows, they like to see light, and so this is particularly important when designing races and crush areas.

We had to eliminate some problems in our system. The cattle would easily walk up the race but always had problems going into the crush. The crush has a roof over it – the problem was that the cattle were going from a race in open daylight into a crush that was dark and often had a lot of shadows. First we put rubber belting on the far side of the race to make it dark as well, and extended the roof to eliminate some of the shadows. We also found that keeping the head bail open allowed cattle to see out the other side, which would encourage them to walk into the crush. Once educated, the latter was no longer necessary but it had helped our more flighty commercial cattle learn to walk into the crush.

Crowding cattle will increase aggression, as their personal space is invaded. Don't overcrowd the yard; half-fill it instead and you will be able to move the stock much faster and more efficiently. Temple Grandin reported that some of the biggest abattoirs in the US have been able to completely eliminate the use of electric stock prods by learning to handle cattle in small groups and not overcrowding them.

Don't yell at cattle and use high-pitched noises, as their hearing is significantly more sensitive than ours in both pitch and range. If you wish to move them, try using a flag on the end of a stick instead. Just waving the stick out to the side, remembering that cattle have excellent 330° vision, is very effective. Banging gates, shouting and yelling all add to their stress levels.

There is no need to hit your animals when working them in the yards. Once again a flag on a stick in front of them will suffice. Continually prodding them can cause them to kick, risking serious injury to the handler.

Cattle look in the direction they are about to move, so by keeping your eyes on their eyes you will be able to predict which way they are about to move. They also like to follow each other, so the sight of a beast in front of them keeps them moving forward. Once you have the mob moving forward then you should stay on the edge of the flight zone and enter it only when cattle stop moving. As soon as they start to move again then you should move out of the flight zone. Problems occur when the handler stays within the flight zone when the cattle have no route ahead, causing them to turn back and break away.

If an animal baulks at a gateway, going into a crush or onto a truck take it calmly. Give the animal time to have a look and assess the situation. Don't rush it.

Cattle can be taught quickly and easily to enter the crush by gentle education. If at first they won't move then tap them on the rear. If the animal refuses after being tapped then twist its tail. The moment it takes one step toward the crush, immediately release the tail. The principle is to reward the animal for moving forward by instantly releasing the pressure. The same principle applies with tapping them on the rear – once they move forward you immediately stop tapping them.

Your cattle will become accustomed to the way in which they are mustered and worked through yards. It's a good idea to work out the best procedures for mustering each paddock and working stock in yards and adhere to those so your animals will learn what is expected of them.

Cattle have excellent memories and will always remember painful or unpleasant experiences. Therefore their first experience in a new set of yards should be a pleasant one. Don't try to tattoo or dehorn them during their first time in a new set of yards or facilities. Allow them to walk through quietly once or twice before catching them in the crush, as this will help reduce their stress levels. If an animal's first experience in the yard is bad then it will always associate the yards with unpleasant experiences. These practices may initially seem time-consuming but they will save you hours and hours over the years.

Other interesting behavioural facts are given below (source: Temple Grandin, http://www.grandin.com).

- It is thought that cattle have only blue and yellow receptors and not the red receptor to see colours.
- Fine cannon boned animals tend to scare more easily.

Tip: Don't try to move or work with stock during the hottest part of the day, especially in the middle of summer. They will be reluctant to move from their shade and if made to do so they can very easily suffer from heat stress. Young calves are especially vulnerable.

- Hair whirls (little twists of hair) found high on the head can indicate bad temperament.
- Hair whirls found low on the head mean a calm temperament.
- Whirls that are wider than the width of the eyes seem to relate to semen quality.

Joining

The term 'joining' is used when females are mated (joined) with a bull.

A heifer's sexual development will cease when she reaches 65% of her mature bodyweight but heifers can start cycling at eight months of age so you may need to take this into consideration when segregating your animals. When a heifer is being joined for the first time, two or three matings may be required before she successfully conceives.

A delicate balance must be struck between healthy and unhealthy weight gain prior to joining. If a female is too fat or thin then she will have problems conceiving and her long-term milk production will be affected. In 90% of commercial herds in Australia heifers do not get to the ideal joining weight and body score before they are joined, which subsequently affects their conception rate.

Q. At what age should we join our heifers?

A. We join our heifers at 16 months of age so that they calve for the first time just after they have turned two. We don't want to join prior to this, as a growing foetus at too young an age will have a negative impact on a young heifer's growth. We also don't want to wait until they have laid down too much fat, which certainly has to be watched in some smaller breeds, as fat deposits around the ovaries can affect their ability to conceive.

Q. Is there any advantage of autumn calving versus spring calving?

A. There are advantages and disadvantages to both. I have both an autumn and spring calving for management and showing purposes and to spread my calving throughout the year. A spring calving means that the cows have to be joined during October, November and December which can, in the hotter areas, lead to heat stress problems for bulls and a subsequent reduced fertility rate. Autumn calving means that cows have to be joined in May, June or July and often results in lactating cows during winter when feed is scarce; in my region, supplementary feeding is necessary. In the southern states, autumn/ winter calving older cows are susceptible to grass tetany that can result in death and spring calving often presents the problem of lack of summer feed.

Q. At what age will our females begin to cycle?

A. This will vary greatly within breeds but some can begin to cycle as early as eight months and others not until they are nearly 16 months. Most heifers will begin cycling at around 10–12 months of age.

Q. How do we know they are cycling? What are the telltale signs?

A. Cows come into oestrus or heat cycle every 18–24 days. The most important stage of the cycle is the period when the cow shows heat. This usually lasts for about 12–18 hours. During this time cows will try to mount each other; the cow that is on heat will stand to be mounted by others. She may lose hair off the top of her tail head from other cows trying to mount her; there may be a clear mucous discharge from the vulva and down her tail; her vulva becomes swollen, wet and flabby; she may exhibit restlessness and bellowing. Some cows are very noisy and display obvious signs while others have no visual signs. These are called 'silent heats'. Sometimes a little blood can be seen, an indication that the cow has ovulated and gone off heat. A small quantity of blood is released by the ovary at ovulation, which occurs after oestrus.

Q. How long after a cow has calved will she cycle again?

A. This is dependent upon the cow's nutritional and lactational state. It normally takes three to four weeks after calving for a cow to return to cycle but the first cycle is normally not as strong, is shorter and of lower fertility. To achieve a calving interval of 12 months cows should be joined two to three months after calving. If the cow is in poor condition then the follicle growth in her ovaries is retarded and she will not cycle again until her body condition improves.

Q. Is it true that if a heifer is weaning her current calf she will have trouble conceiving?

A. The stress of having a calf taken away could cause a heifer not to conceive if she happens to be cycling at the time. This will only last for a week to 10 days. Don't ever wean a calf during an AI or embryo transfer program, as the stress could definitely affect the conception rates.

Q. At what age will a bull work, i.e. get cows in calf?

A. This will depend on the maturity pattern of the bull but a number have been known to do the job very successfully at eight months of age! By 12 months they will certainly be able to work – not to the full capacity (i.e. large numbers) of a two-year-old bull but for small numbers they will work perfectly.

Q. How many cows can a bull be joined with?

A. This is called a bull's serving capacity.

- One yearling bull to 15–20 females.
- One two-year-old bull to 25–35 females.
- One mature bull to 30–50 females.

Q. When is the best time to put the bull in with the female?

A. Nature's way is best: a bull knows when a female is ready to be mounted and joined by smelling her and observing her behaviour. This is called 'standing heat', when the female stands still to let the bull mount her. It usually occurs some 10–18 hours after the cycle begins. Females are normally left to run with a bull for a period of three months.

Q. I am currently running Santa Gertrudis cattle on my property and am looking to join my heifers much earlier than usual, assuming they reach the target weights and are cycling properly. I was wondering about the potential of a Lowline bull to guarantee low birth weights and easier calving for the heifers? What weight are the calves normally born at?

A. Any of the small breeds will guarantee lower birth weights. Obviously it will depend on the frame size of the cows you are joining the bull to. Fullblood Lowlines have normal birth weights of 18–25 kg. When I join them to my commercial cows then my birth weights are 25–35 kg, depending on the breed of the cow.

Q. This must be a first, but I need to know if there is anything I can do to either stop a heifer coming into heat or if there's something to put on her to keep a bull off her.

Table 8.2: Cattle gestation table

Date of service	Date of birth	Date of service	Date of birth	Date of service	Date of birth
1 Jan.	12 Oct.	7 May	15 Feb.	10 September	21 June
8 Jan.	19 Oct.	14 May	22 Feb.	17 September	28 June
15 Jan.	26 Oct.	21 May	1 March	24 September	5 July
22 Jan.	2 Nov.	28 May	8 March	1 October	12 July
29 Jan.	9 Nov.	4 June	15 March	8 October	19 July
5 Feb.	16 Nov.	11 June	22 March	15 October	26 July
12 Feb.	23 Nov.	18 June	29 March	22 October	2 August
19 Feb.	30 Nov.	25 June	5 April	29 October	9 August
26 Feb.	7 Dec.	2 July	12 April	5 November	16 August
5 March	14 Dec.	9 July	19 April	12 November	23 August
12 March	21 Dec.	16 July	26 April	19 November	30 August
19 March	28 Dec.	23 July	3 May	26 November	6 Sept.
26 March	4 Jan.	30 July	10 May	3 December	13 Sept.
2 April	11 Jan.	6 August	17 May	10 December	20 Sept.
9 April	18 Jan.	13 August	24 May	17 December	27 Sept.
16 April	25 Jan.	20 August	31 May	24 December	4 October
23 April	1 Feb.	27 August	7 June		
30 April	8 Feb.	3 September	14 June		

Birth dates based on 285 days.

A. No, there is nothing that can be done. You can give the heifer 2 mL of PG 10 days afterwards if you want to abort any pregnancy.

Q. How long is gestation?

A. This is the period from conception to birth of the calf. It can vary slightly between breeds but usually averages 280–285 days (Table 8.2). We have found our Lowlines have a gestation period of up to 14 days less, i.e. 271 days.

Calving

This is a special time of the year, especially for small breeds due to the size and cuteness of the calves. A number of commercial producers have noted how much more active the small breed calves seem to be. If you observe a paddock full of young calves, especially in the late afternoons, you will often see them playing and chasing each other around. It is important for calves to be run together so they have that interaction and learn some social skills. They also learn to define their hierarchy at this very early age.

A calf normally gets up within 30 minutes of being born and has its first drink within one hour. It is vital that a newborn calf drinks from its mother within 24 hours as this milk (known as colostrum) is extremely rich in vitamins and antibodies (five times higher than that of normal milk). After 24 hours the cow stops making colostrum and starts producing milk. Calves that do not receive adequate colostrum will be stunted and often affected by disease due to their reduced immunity. After three months the immunity offered by the colostrum begins to fade and their own immunity kicks in.

When a calf sucks milk from its mother's teats it kinks its head in an upward position which nature has devised for a very specific reason. This allows the oesophageal groove to form, which causes the milk to bypass the rumen and go directly to the second stomach. The rumen contains a bacterium that, although necessary to break down pasture and fibre, feeds on proteins including milk proteins. The suckling process allows milk to bypass these bacteria, making all proteins available to the calf.

Remember to check cows that are calving at least once a day. We check ours twice a day, especially when it is heifers that are calving.

Q. How do I know when a cow is about to calve?

A. There are a number of signs to look for.

- First her udder will begin to fill, also referred to as 'bagging up'. Just prior to calving the udder will be very tight and full, and sometimes milk can be seen dripping out of the teats. Heifers, or first-time calvers, usually don't bag up nearly as much so this can be difficult to detect. This period is also referred to as 'springing'.
- A cow will become very loose and soft around the vulva.
- Sometimes the tail head will rise slightly.
- You will notice a mucous vaginal discharge as the birth canal is lubricated.
- The cow will become restless and usually go off on her own to have the calf away from the rest of the herd. If you see a very pregnant cow off by herself, that is a sure sign that the birth is imminent.

Q. How do I know if I need to call the vet?

A. One of the advantages of small breeds is that they have relatively few calving problems due to their small birth weights. In 20 years of breeding Ausline and Lowline I have only ever lost one calf (due to a breech birth) and had to assist two heifers with their calves.

This heifer is very close to calving. She is very loose around her vulva, has a discharge and her udder has milk.

There are a number of signs to look for. If the water bag has broken and there is no sign of a calf within two hours of straining, the cow will need assistance. The quicker assistance is sought, the greater the chance of the calf surviving – once the water bag has burst the calf will be viable only for four to six hours. If the nose and front feet are visible, then a gentle pull on the calf's feet in time with the cow's contractions may be all the help that is needed. If the cow is quiet, you may be able to do this in the paddock. If, on the other hand, two rear feet are visible or you cannot get near the cow then she will need to be brought into the yards quietly and put in the crush so the vet can assist. With a posterior presentation, time is of the essence. Once the hind feet are showing, the cord is then pinched and the unborn foetus – with or without its blood supply or the opportunity to breathe – will quickly suffocate due to the position of its head.

If a heifer has had a difficult time pushing and expelling the calf, she may be too tired to clean the calf after its birth. Assistance maybe required to stimulate the calf's breathing by vigorously rubbing its chest and cleaning away all mucus and blood from its nostrils and mouth. Hanging the calf

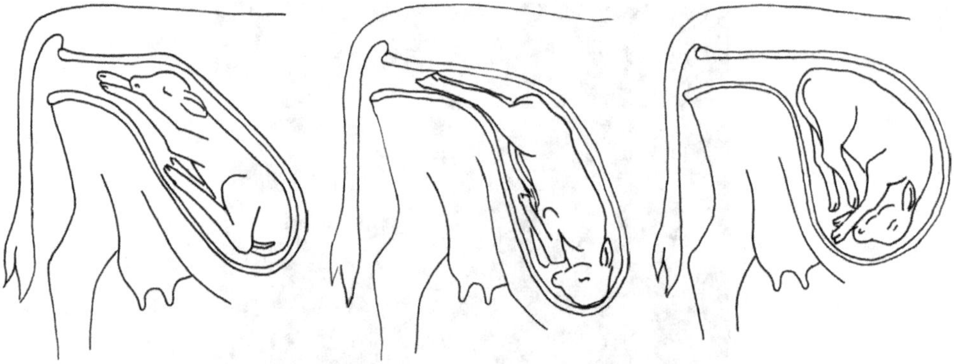

Normal presentation (left), normal posterior position (middle) and breech position (right).

upside down, by putting it over a gate or railing fence, to expel any mucus from its lungs can also work.

What to do if a cow rejects her calf

Heifers, and cows for that matter, sometimes feign disinterest in their calf or reject it altogether, due to inexperience or a long or difficult birth. If possible – and you will need perseverance – try to get the cow to accept the calf as it's a whole lot easier than bottle-feeding, which is time-consuming. I don't think bottle-fed poddy calves ever do as well as if they are on mum. Try to avoid handling the calf too much and covering it with your scent as this will only confuse the cow. Some strategies are discussed below.

- It is advisable to keep the cow and calf in close proximity for a few days, such as in a yard, until you are sure the calf is sucking. Sometimes the hormonal triggers are slow to kick in but the sight and sounds of her calf will assist.
- You may need to put the cow in the crush every four hours and restrain her while you teach the calf to suck. Once it has developed a strong sucking action it will normally be able to drink from mum without any assistance. This is where perseverance is often required. Once the calf has learnt to do this, then often it will be strong enough to convince mum to accept it and let it feed.
- Some cows will kick and head-butt their calves, and you may need to intervene to prevent injury to the calf. Restrain the cow while the calf drinks. Sometimes it takes only one feed for her to change her mind but it may require three to four days.
- You may also need to milk the cow and bottle-feed this to the calf until it is strong enough to suckle on its own, especially if the calf has been rejected for some time.

U-Bond is a masking agent that aids in establishing a bond between calf and mother if the calf has been rejected. It can be bought online and from most produce stores.

One of my cows calving, that I sold to Platinum Park Auslines. The bag appears then the cow lies down to push out the calf. You can see the feet and nose appearing before the whole body is expelled. The cow immediately begins to lick the calf and clean it. Very shortly it is up and looking for its first drink. (Photos courtesy of Platinum Park Auslines)

I once had an Angus heifer that gave birth to a calf. As soon as the calf was born she tried to kill it; she kept lifting it up in the air and throwing it. I rescued the calf, which was in a dangerous situation, and brought them into the yards. I put the calf in a separate yard beside the heifer. After a couple of hours she calmed down and I eventually got her to let the calf suckle. She began to mother it.

I also had a big commercial Limousin cow that, after six births without problems, rejected her last two calves. It was obvious that as she was getting older she must have suffered some hormone imbalance. We did get her to accept both calves by locking her in the yards with them for a week or so. We also put her in the crush four times a day to teach the calf to suckle, but it was very time-consuming . We decided to eventually get rid of her.

Q. Is there any special care I need to provide for a newborn calf? I have read that I need to make sure they have their first drink.

A. Very rarely do newborns need extra attention. It is a good idea to make sure that, within 24 hours, they have their first drink of colostrum. Just by observing from the edge of the paddock you will be able to see if a calf is sucking. Once it has had that first drink, subsequent problems are very rare. We store colostrum, which we milk from our very quiet cows, in the freezer in case of emergencies. If you're not handy with milking a cow there is a great hand-held machine called an Udderly EZ milker that makes it a very fast and efficient process and excellent for making sure you have some colostrum. You can buy colostrum in powder form (under the brand name Impact) and we always have that on hand as well.

If, for any reason, you do have to move a cow and her newborn try to make sure she has had a chance to smell the calf. If you move a calf from the birth site, make certain that the cow follows and that she can see you and the calf. Don't totally dry a newborn, because birth fluids are a strong stimulus in eliciting maternal behaviour.

The masking agent U-Bond.

Q. We lost a cow due to snake bite (we think) and were wondering if it is possible to foster her calf onto another cow.

A. This will depend on the age of the calf and the cow you are trying to foster her onto. Dairy cows often make the best foster mums as some will

Tip: A cow's milk production is always lighter for its first calf. That is why a lot of commercial producers never keep the progeny from first-time calvers. Cows don't normally come into full milk production until their second or third calf.

take up to two or three additional calves and have the milk supply to feed them. Fostering is not always an easy process and can be time-consuming. The problem is due to the fact that each calf has its own scent and cows recognise calves that are not their own. To overcome this you can try using U-Bond or putting vanilla essence on the calf to be fostered, as this will hide its scent and confuse the cow. Otherwise you will need to put the cow in the crush and restrain her in a head bail. Open the sides of the crush and give the calf access to the udder, allowing it to drink. The cow will probably try to kick the calf, so sometimes giving it access from behind the cow is easier. Or you can hobble the cow. You will need to do this three to four times a day to ensure the calf receives enough milk. It may take days for the cow to accept the calf and sometimes they never do accept a foster calf.

Q. Can I have breeders if I don't live there? I live in Canberra. I can visit the farm on weekends but I have uni during the week. I thought I always had to be with them in the weeks around when they were expected to give birth in case they needed help with the delivery.

A. It is possible to have breeders if you don't reside on your property. One of my clients has commercial cows and uses Ausline bulls over all his cows. He does not reside on his farm. He goes there only on the occasional weekend and he has never had any calving issue. This is because he is using a small breed bull, which will guarantee a low birth weight and hence reduce the incidence of calving issues. His females are all on their second and third calf. It would not be an advisable option with stud heifers having their first calf.

Q. How do you tell when a calf is born if it's a boy or a girl?

A. It can be difficult to tell until they are a couple of days old and the umbilical cord has fallen off. If you can get close enough, put your hand between its legs and see if you can feel or see the two testicles. Alternatively, observe and watch them urinate.

Q. My miniature Lowline was born three weeks ago and was half the size of the other eight calves we have had over the years. We think it arrived three weeks early and the mother rejected it as it didn't stand for 24 hours. I have been caring for it. He is going OK but I am at a loss as to how much to feed him. I had been giving him small feeds every two hours in the first week and then four-hourly. I now use 150 g milk powder to 1 L warm water, three times a day. The calf is growing but still having trouble with his legs. Some days he is running about then the next day can hardly get

up. His poo and wees are good. I have him in a shed and he wears a little coat I made for him. Any advice on what else I can do? This is the fifth calf from this mother and the first one to have troubles.

A. My advice is to give him as much as he will drink. Use a high-quality milk powder specially designed as a milk replacer for calves with a protein content of 18–25% and a fat content of 10–20% (the higher the better). Dairy farmers know the best brands. The brand will give you the dilution rates. Your local produce store will sell milk powder. Also make sure he got some colostrum from his mother. Do you vaccinate for pestivirus? Pestivirus can cause cows to abort or have premature calves and calves born with a host of issues such as coughing, inability to walk, blindness etc.

Weaning

Q. What is weaning?

A. This is the process of permanently removing the calf from its mother so it can no longer suckle.

Q. At what age should we wean the calves?

A. We wean between six and 10 months depending on a number of factors.

- The season and the condition of the cows: if the cow seems to be suffering and losing condition due to an exceptionally large calf or lack of pasture, then we would wean sooner rather than later. The reason is that there is no point keeping the calf on its mum as not only will the cow be declining in condition and therefore take a lot longer to regain weight but the calf is probably also suffering due to an inadequate milk supply. It is also a cheaper option to wean the calf and supplement-feed it than try to feed both cow and calf. The cow will quickly regain her condition once she is no longer required to feed a calf.
- Type of production: if you are producing and selling vealers straight off their mums then you may want to wean later.
- Heifer calves: you don't want your heifer calves to get too fat at puberty as this can affect their future milk production. Laying down fat in the udder affects the development of the milk-secreting tissue.

  Remember that most small breeds are early-maturing cattle and some young bulls can work from eight months, so keep an eye on them or steer them early.

Keeping in mind that we aim for a cow to calve annually, then it is important that a pregnant cow not suckle a calf during her last trimester. She needs to be able to put all her energy into a growing foetus, not into both a growing foetus and a milk-sapping calf. Be careful. You will be surprised at how a cow will take her calf back even after two months of absence.

Q. What is the best way to wean calves?

A. There are a number of methods of weaning.

- **Yard weaning**: we find this to be the best way for two reasons, in that it's fairly stress-free and it also serves as an educational process for the calves. The calves are put in the yards and the cows are run in an adjoining paddock or yard for a number of days so they can see, hear and smell their calves. The calves are fed good-quality hay and weaner pellets on a daily basis for at least three weeks. This prevents them from losing excessive weight and helps them get used to human interaction. After a couple of days both cows and calves will stop bellowing and you will be able to move the cows back to their paddocks. If you try to do this on day one you will only end up with very distressed cows and calves and probably a lot of broken fences. A cow naturally wants to protect her calf and will do everything in her power to get back to her calf. A calf will do everything in its power to get back to drinking mum's milk.
- **Total separation**: this is quite common on larger properties. The calves are drafted from their mothers and moved as far away as possible. This is very stressful for both cows and calves.
- **Paddock weaning**: the cows and calves are placed in adjoining paddocks. The cows move off to graze but can still see their calves. After about four days they can be moved further away. The issue is that some calves can still drink through the fence. If so, they need to be shown where their water supply is or they will remain at the fence all day waiting for the return of their mothers.
- **Creep weaning**: this is a self-weaning process in which the calves have access to a ration or quality pasture through a creep gate that allows them through but not their mothers. The calves become accustomed to grazing without their mothers and eventually they can be shut off permanently with a lot less stress.
- **EasyWean®**: these are specially designed nose rings and are an excellent alternative for those who have a small number of animals and limited space to separate weaners. The nose rings are metal or plastic and fit

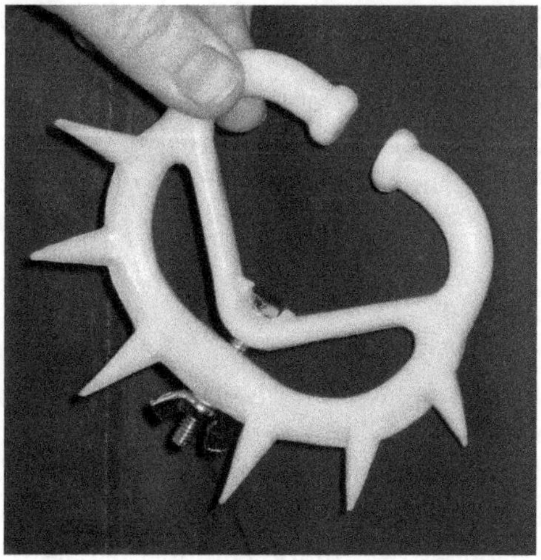

EasyWean® nose ring.

into the calf's nose, with a number of points protruding. They don't cause pain to the calf, but each time it tries to drink from mum it causes discomfort so the mother will not allow the calf to suckle. Eventually the calf will give up and mum's milk production will cease. EasyWean® has a number of advantages in that the cows and calves can continue to run together, stress levels are reduced and the rings are reusable.

Q. We have a three-year-old cow that drinks milk from other cows. What can we do?

A. My advice is to get rid of this cow or put her in a paddock with heifers. She was never weaned correctly. She was probably never taken off her mother at all. Her mother probably weaned this cow herself when her next calf was born. You do not want animals like this in your herd as they will continue drinking all the milk from any lactating cow for the rest of their life, thus depriving calves of this essential food source.

Vaccinations

The importance

Herd health is an important factor as loss from disease can be very expensive, especially when veterinary treatment is needed. Many diseases are highly contagious and can affect growth and fertility. Preventative herd health is a very cheap alternative.

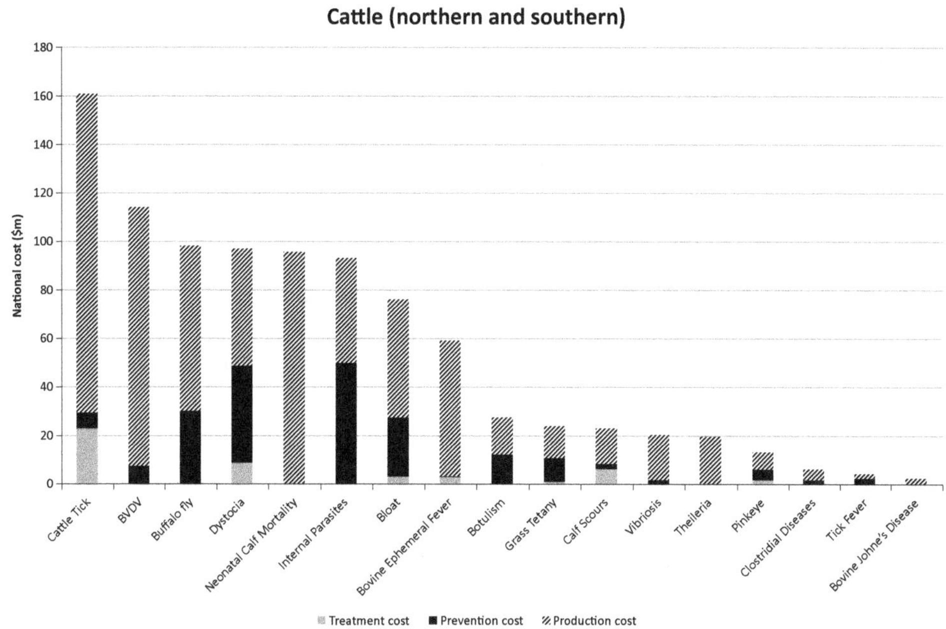

Estimated annual economic cost of the priority diseases for cattle. Source: Modified from 'Priority list of endemic diseases for the the red meat industries' (2015). Report prepared by GHD Pty Ltd for Meat & Livestock Australia Limited.

Types of vaccines

Killed vaccines usually require two initial injections at least four to six weeks apart for maximum effect. An annual booster is required to sustain protective immunity. Killed vaccines are a mix of the disease (dead) and compounds called adjuvant that stimulate the development of immunity. Water-soluble adjuvants are preferred, but sometimes oily adjuvants are used to get enough stimulation; examples include Vibrovax®. The extra stimulation can also cause extra and prolonged site reactions if not administered properly.

Live vaccines, such as those for tick fever, have altered organisms to produce immunity but not disease. Most require one injection.

Reactions

All vaccines cause significant reactions and pain for up to a week, to the point of lameness in the odd few even when given properly. A swelling can occur at the injection site. Severe reactions are rare, but if one does occur seek veterinary assistance. If practical, it is advisable to avoid using more than two bacterial vaccines at once.

Vibriosis may cause temporary subfertility in bulls so vaccinate two months prior to mating.

Technique

Always keep your vaccines in the fridge. Buy pack sizes to suit your needs and check their expiry date to ensure that they will last long enough for you to use it all. Keep the vaccine cool while it is being used, and out of the sun. Always check the dosage rate as this can vary between manufacturers, especially for 7-in-1.

- Use short (12 mm) and sturdy (18 gauge) needles. Inject only under the skin, not into the muscle. Insert the needle as gently as possible.
- Avoid vaccinating wet cattle, as the chance of infection at the injection site is much greater.
- Avoid injecting more than one vaccine into the same site. Try to keep injection sites at least one hand-width apart.
- The objective when vaccinating is to get the opening of the needle resting between the skin and underlying tissues. Entry should be at about 45° to the skin, with the bevel parallel with the skin.
- Most vaccines for cattle should be given under the skin, i.e. subcutaneously, especially oil-based vaccines. If the vaccine is given into muscle, severe reactions can occur.
- Hygiene is important. Vaccinator guns should be disassembled, cleaned, sterilised and reassembled between each use. Disposable needles should be put in a sharps container.

Recommended vaccines

7-in-1 is a vaccine that prevents clostridial diseases and leptospirosis and reduces shedding of leptospires in urine. These diseases cause significant productivity losses every year.

Clostridial diseases include blackleg, tetanus, malignant oedema, pulpy kidney and black disease. They are among the most common causes of death of cattle around Australia. Unvaccinated stock contracting any of these five diseases have very little chance of recovery.

Leptospirosis results in reduced weight gain, reduced milk production and in some cases death. More importantly, lepto can be transferred from cattle to humans, resulting in health and legal implications.

5-in-1 is a vaccine that protects against all five clostridial diseases: tetanus, malignant oedema, blackleg, enterotoxaemia and black disease. 5-in-1 can also be used on sheep whereas 7-in-1 cannot.

Pestivirus

Pestivirus is also known as bovine viral diarrhoea (BVD). BVD is one of the most costly diseases affecting Australian beef production with 70% of Australian herds showing evidence of past BVD infection. It can result in reduced conception rates, abortions and birth defects. It is transmitted by persistently infected animals, also known as shedders, through any mucous membrane and is highly infectious. When a dam is infected during pregnancy, the virus can cross the placenta and infect the developing foetus, often causing the animal to abort. If the foetus survives it will usually be born with abnormalities such as weakness, inability to suckle, blindness or neurological problems (wobbly gait, tremors), and die soon after birth. If it does survive it may be a shedder. Those that are persistently infected often don't grow properly and suffer other illness such as calf scours or pneumonia. They usually die before 18 months of age from mucosal disease or respiratory disease but some can appear 'normal' and act as long-term carriers.

I learnt the hard way when it comes to pestivirus. I knew nothing of the disease until my cattle were housed next to infected ones at a show I attended in 2000. Some six months later my very best show cows starting giving birth to pestivirus-infected calves, which all eventually died. By that time even more damage had been done, as these shedder calves had infected a number of other pregnant females and so the cycle continued. It took two years to eradicate the disease from my herd and I lost a significant number of progeny in the interim. Since then I have vaccinated every year.

That's why it is important to obtain health records for new stock you've purchased. If these are unavailable, then isolate new stock for 30 days until vaccinations can be administered and worming treatment can be applied. Tests for this are some of the simplest yet most accurate and are often required prior to entering a number of royal and local shows.

More information is available at https://www.bvdvaustralia.com.au.

Tick fever

Tick fever vaccination (also referred to as 'blooding') is necessary for those who reside in a tick area, otherwise you run a very serious risk of your cattle dying. Tick fever or red water is carried by the cattle tick, and can result in death, loss of condition, sterility in bulls and abortion. Fever and red urine are typical symptoms of red water, while anaemia, jaundice, weakness, loss of appetite and condition are common to tick fever in general.

The once-only highly effective vaccine is available from the Queensland Tick Fever Centre and provides life-long immunity. It is a live vaccine and has a shelf life of only four days. The ideal age to vaccinate is calves between three and nine months.

Some producers rely on natural immunity, but this strategy is risky and not one I would recommend. Cattle born and raised in areas where cattle ticks are endemic can develop natural immunity through exposure to ticks infected with tick fever. Cows with tick fever immunity provide their calves with a natural immunity that lasts approximately for the first three months. The tick fever website states the following: *Most calves develop an age-related resistance that protects them from about 3 months until about 9 months of age. This is irrespective of breed and whether dams are immune. Calves exposed to tick fever when the age-related resistance is present rarely show clinical signs and develop a solid, long-lasting immunity. If immunity develops in all your calves, tick fever will not be a problem, but they must be exposed to all 3 parasites. This requires exposure to thousands of ticks because not all ticks carry the infection. The age-related resistance gradually wanes from 9 months of age. If they haven't been exposed to and developed immunity to tick fever, they will become highly susceptible to tick fever. If then exposed to infected ticks later in life, these cattle are likely to develop a severe, life-threatening infection.*[2]

If cattle are introduced from a tick-free to a tick-prevalent environment they will have no immunity and will certainly die from a tick infestation. Ideally, cattle introduced from tick-free country should be vaccinated at least two months prior to relocation to avoid the risk of field infections if exposed to cattle ticks prior to developing immunity. If this is not practical, they should be kept tick-free for four weeks following vaccinations. The tick fever vaccine can cause reactions in some cattle and it is advisable to monitor animals older than nine months, bulls and pregnant females.

Cattle suffering from tick fever can be successfully treated if they are diagnosed early enough. Imidocarb (Imizol® or Imidox®) is the best drug for this purpose. Oxytetracycline antibiotic can be used to treat anaplasmosis.

Bovine ephemeral fever (BEF – three-day sickness)

BEF is a viral infection in cattle that is characterised by an acute onset of lameness, depression and fever. Most cattle will recover. The virus is specific for cattle, and starts at the beginning of the wet season in northern Australia then spreads south and east down the east coast. It then spreads into southern Queensland, and central and coastal New South Wales. Occasionally, as happened in 2010, outbreaks occur in southern New South Wales and northern Victoria.

It is very contagious as, once an infected beast is bitten by midges and mosquitoes, the infected blood can be transmitted to other animals as the insects move about the herd. The virus specially attacks joint tissue. Cattle usually develop an acute fever with their body temperatures rising to 40°C, although the fever may fluctuate. The muscles become sore and stiff and the cattle become lame. Respiratory rates and heart rates increase and a nasal discharge may be seen. The cattle may lie down for two to 10 days, but the average is three to five days. Milk production in dairy cattle falls rapidly and pregnant cows may abort. Semen quality in bulls is severely affected due to the fever and may take up to three months to recover fully. Once exposed to the virus, cattle develop a long-lasting immunity.

The only method of treating affected cattle is to bring food and water to them regularly and provide some sort of shade if necessary. Anti-inflammatory drugs can be given to relieve muscle stiffness, decrease their temperatures and make them feel not so miserable. Some of my clients have used ice packs to reduce exceptionally high temperatures.

To date we have never had an animal die from three-day sickness but it can be time-consuming if you have a number of animals affected at the same time. There have been some problems with the vaccine in recent years as the virus is mutating so quickly that the laboratories have had problems keeping up. This has resulted in some vaccinated animals showing minor symptoms of the illness, but in my experience they have been affected far less severely – the disease has only made them lame and sore.

Vibriosis

Vibriosis has been called 'the quiet profit-taker' and is widespread throughout Australia. It is a venereal disease caused by a bacterial infection that can significantly reduce herd fertility and reproductive performance without distinctive or obvious signs. Vibriosis is spread at joining either from an infected bull to an uninfected cow or vice versa. Infected bulls can be carriers for many years and are therefore the main source of infection in a herd. Clinical signs include low pregnancy rates: sporadic abortions are usually the initial indication of a vibrio-infection in the herd. Infected herds often have conception rates below 50%. Vibrio infection results in reproductive damage and can cause permanent infertility in up to 11% of heifers. Infected bulls show no obvious signs but can act as long-term carriers, infecting cows and heifers at service.

Pinkeye

Pinkeye (infectious bovine keratoconjunctivitis) is an inflammatory bacterial infection of the eye that can cause permanent blindness in severe cases. It is a contagious disease that occurs mainly in young cattle in summer.

Pinkeye can affect up to 80% of a mob, with affected weaner calves losing up to 10% of their bodyweight. Although rare, deaths may occur when both eyes are affected, due to starvation, thirst or misadventure.

Infective material discharged from the eyes of affected cattle can be spread to other animals by flies, or onto long grass grazed by the cattle. Sunlight and dust make the problem worse and outbreaks are often worse in very dry conditions.

The first signs seen are:

- copious watery eye discharge;
- aversion to sunlight;
- signs of irritation, e.g. excessive blinking;
- reddening and swelling of the eyelids and the third eyelid.

The eye will then go cloudy in the middle and the cornea may ulcerate during the next two days. If left untreated the animal can go blind. Stock under two years

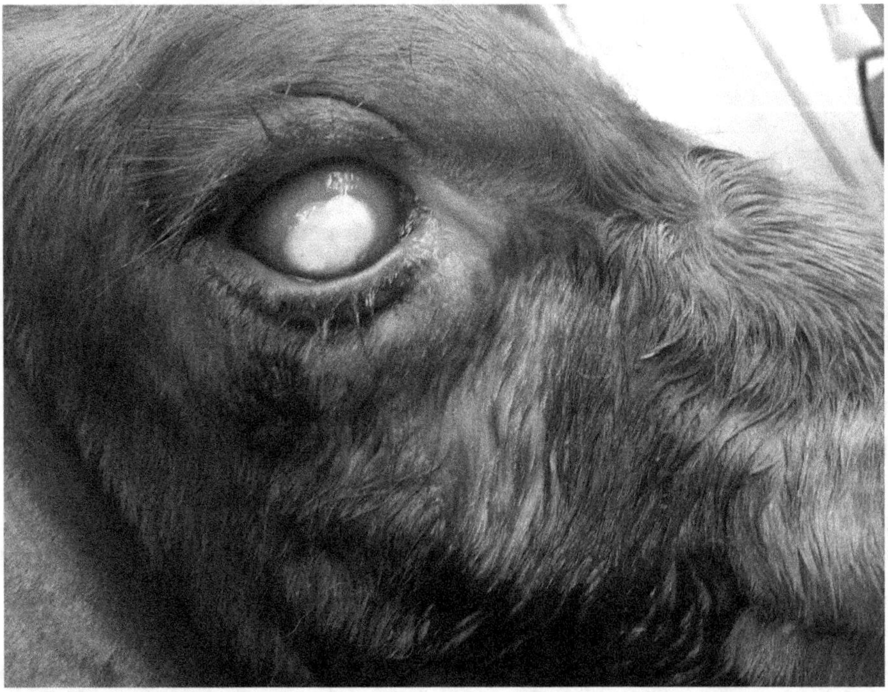

This animal has an extremely affected eye which has ulcerated.

are highly susceptible. The optimum time to vaccinate is about October, before the flies becomes too dense. This gives six months of coverage.

There are a number of treatment options.

- Several proprietary formulations of eye ointments available under prescription from veterinarians, with the active antibiotic constituent cloxicillin: cloxicillin-based eye ointment is applied into the conjunctival sac (under the upper and lower eyelids). A single application is usually sufficient, but it may be repeated every 48 hours in severe infections. I use Opticlox and find it very effective. I treat any infected animals twice a day for three days or even if I see the first signs of an infection. The secret is to treat them early to prevent permanent scarring.
- Aerosol sprays: teramycin-based product in a very convenient spray can that can allow you to treat quiet animals in the paddock.
- Eye patches: these can be glued over the infected eye once it has been treated with a long-acting eye ointment. They prevent dust, flies and sunlight further irritating the animal.
- Antibiotics injected directly into the eye: this is the Rolls Royce of treatment and will require a vet to administer. If you have an animal that has a severe infection and you don't want permanent scarring, this method is superior.
- Vaccination is a preventative measure which helps reduce the incidence.

Q fever

Q fever is the most common disease that humans catch from animals and the main carriers of the disease are farm animals such as cattle, sheep and goats. In rural areas, kangaroos can also be carriers. Humans usually become infected through breathing in small particles with bacteria from animal body fluids such as blood, urine, faeces and birthing fluids. Handling birthing products and slaughtering animals are particularly high-risk activities. The organisms which spread this disease are normally contracted from the placental tissues and fluids so if you are assisting in births, handling newborns or touching afterbirth then you are most at risk. An Australian Q fever vaccine is available and recommended for people who work regularly with or are at risk of exposure from potentially infectious animals or materials.

Q. What vaccinations do you give your cattle?

A. I am pedantic about vaccinating as I consider that skipping vaccines is not worth the risk. I won't buy cattle that don't have vaccination records. All our cattle are vaccinated for:

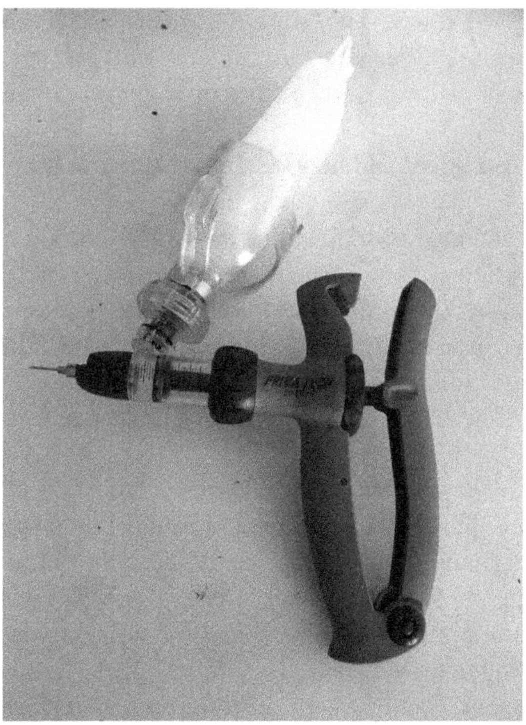

Vaccinator guns are excellent as the amount of vaccine to be administered can be set (there is no wastage). The short needle ensures the injection is subcutaneous.

- **tick fever**: once-only lifetime vaccination of 2 mL. Immunity can take up to four weeks. Monitor animals over nine months old for reactions;
- **7-in-1**: two shots initially (four to six weeks apart) then an annual booster;
- **three-day sickness**: two shots initially (two to four weeks apart) then an annual booster. Only for cattle older than six months. Vaccinate before spring to ensure full immunity;
- **vibrio** (for bulls only): two shots initially (four weeks apart) then an annual booster;
- **pestivirus:** (females only): two shots initially of 2 mL then an annual booster. The second dose can be given up to six months later. Safe on pregnant cows. Immunity does not develop until 14 days after second dose;
- **pinkeye:** single dose of 2 mL annually three to six weeks prior to the onset of the expected pinkeye season. Calves as young as one week can be vaccinated and we vaccinate all our stud animals in October;
- **multimin:** given to all cows and calves at weaning.

Q. When is it best to give 7-in-1?

A. Our calves are given their first shot one month prior to weaning and the second shot is given at weaning. Our cows are given their annual booster in September. It is wise to vaccinate bull calves at least one month prior to castration, to allow some immunity to build up to prevent tetanus.

Q. Where do you inject and how far does the needle have to go in?

A. Most vaccinations are given subcutaneously (just under the skin, not into the muscle). The skin is pinched up to ensure that the injection does not go into the muscle. You need to be careful because if you do go into the muscle you can cause an abscess; using short needles will help prevent this. If you are selling animals for slaughter you will be penalised if there is needle damage on the body, hence injecting in the neck is recommended. There are two options:

- individual disposable syringes and needles;
- vaccinator guns, which can use either short or long needles. Just make sure you change your needles regularly … the manufacturers recommend every 100 injections but we change them every 50. We use bottle-mount vaccinators for speed and efficiency as they save time by not having to draw up each individual dose.

The recommended site is high on the neck. We also use the area just behind the shoulder where there is plenty of loose skin, but again be careful that the injection is only subcutaneous.

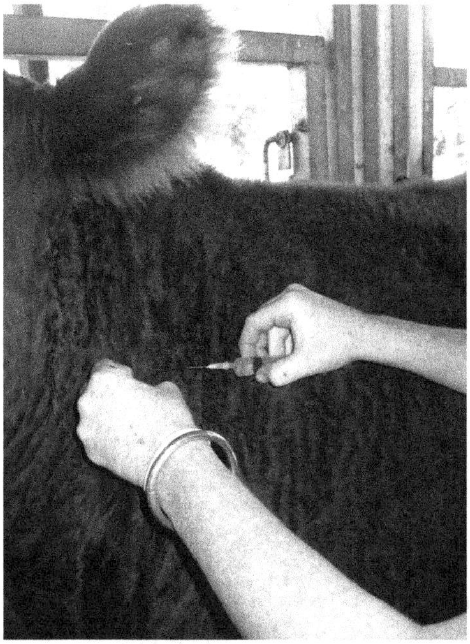

Injection site for vaccinations.

Q. When is it best to vaccinate for vibrio?

A. One month prior to joining. We vaccinate all bulls at working age.

Q. When should you vaccinate for tick fever?

A. Calves have certain immunity against tick fever up until six months. The Department of Primary Industries recommends all calves be vaccinated between three and six months. The vaccination lasts a lifetime. Older cattle can also be vaccinated but the risk of a reaction increases with age.

Q. Can pregnant cows be given tick fever injections?

A. No, not in the last trimester as it can cause them to abort. You will have to monitor the cattle **very** closely until they have been vaccinated as tick fever will kill cattle very quickly. You will need to keep them clean of ticks until they have calved. Cattle that have come from a tick-free environment are particularly susceptible and need to be monitored closely until the vaccination has taken effect (six weeks after injection).

Q. Which is the best month to vaccinate for three-day sickness?

A. One month prior to the start of the three-day season in your area. In my area of south-east Queensland, this is normally February.

Q. Can pregnant cows be given three-day injections?

A. Yes.

Q. How do you handle three-month-old calves when you have to give them a needle? I tried to halter them and push them against a fence but I still found them too hard to handle for me to inject safely. I failed totally and did not want to persevere with other options like the crush at that time as I had already stressed them. I thought they seemed too small to put in the crush and head bail them.

A. There are are few options:

- put a few into the crush together so they have little room to move, then hop in with them and lean against each one as you vaccinate;
- put them in the head bail, with a parallel squeeze which prevents them from moving too much and then inject them;
- lean against them either up against a yard panel or in the race. You will need a second person as, once you have them restrained, you will need someone to inject them quickly;
- use a calf cradle.

Q. We are having difficulty injecting our young calves and we were wondering if a calf cradle would make injecting easier and minimise risk of damage. Have you ever used a calf cradle for purposes like that and was it successful?

A. I have over 150 head of cattle but I don't use a calf cradle. Calf cradles do work well but as we have a parallel squeeze we use this instead for the larger calves. We find the best way for the smaller calves is to get in the crush with them and push them up against the side to hold them. You will require two people, as while one is immobilising the calf the other will have to vaccinate.

Q. Our neighbour's bull got in and serviced one of our best stud cows. Could you please tell me the name of the stuff you inject heifers/cows with to induce them to abort? I assume you get it from a vet.

A. The product is called Lutalyse and is given intramuscularly. If the wrong bull has serviced a cow then Lutalyse should be administered 10 days after

Tip: If you want to collect semen or embryos from an animal for export be careful what vaccinations are administered and when. Check the protocol requirements with your vet or AQIS before vaccinating, as some will give a positive test which may make qualifying for export either impossible or a much more expensive process.

the joining date. It is not to be given to cows in their second or third trimester. Only a vet can only dispense this product.

External parasites

These include lice, ticks and buffalo fly. There are a number of ways to control these parasites and by far the easiest for a small number is applying a pour-on. There is a huge variety of pour-ons and many will control all parasites at once. Dose rates are usually based on weights, so a set of scales ensures accuracy and avoids wastage. The effective control period will vary with the product, ranging from 21 days to 42 days. Do your research prior to purchasing a product as some are based on 1 mL per 50 kg and others on 1 mL per 100 kg. Also compare the effective control periods. The cheapest products often last the shortest time and in the long run are not really the cheapest as you will be treating your animals far more often. Other methods of application include sprays and ear tags.

As the majority of pour-ons are systemic there is no need to treat young suckling calves as the active ingredient is passed to the calf through the mother's milk.

Ticks

Cattle ticks are a serious economic problem for the Queensland cattle industry. The effect of tick infestation is mainly lost production due to reduced growth. Heavy cattle tick infestation causes loss of condition and even death because of tick-worry and blood loss. Hides of heavily infested animals are damaged by tick bites, which reduces their value. In severe cases hides may be unsalable. Other ticks commonly found on cattle in Queensland are scrub ticks and New Zealand cattle (or bush) ticks. Cattle are the main hosts for cattle ticks although they may also be found on horses, goats, sheep, deer, camels and buffalo.

Cattle should be vaccinated for tick fever between three and nine months of age, with a one-off lifetime vaccination. The vaccination will prevent an animal dying from tick fever but it does not prevent it being infected with ticks.

Sprays, pour-ons and injectables can control ticks. Some of the products used to control ticks not only have a worm mixture added but are also able to inhibit a

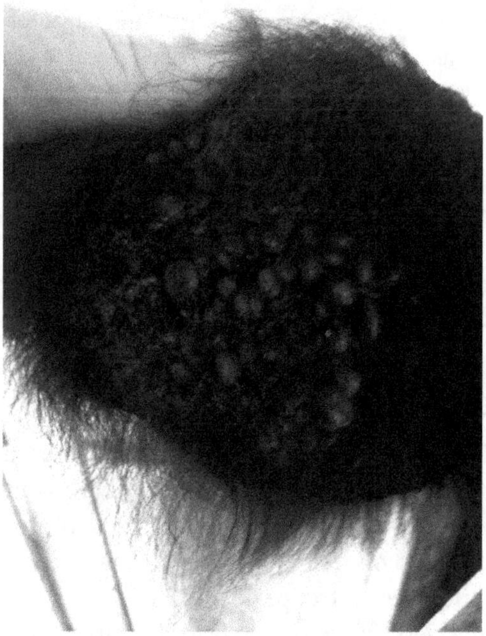

An ear that is heavily infested with ticks.

tick's development as growth regulators. The systemic action of fluazuron breaks the tick's life-cycle at several stages, resulting in sustained tick control on the animal and pasture. While these products are more expensive to purchase they are cheaper in the long term as they reduce overall tick numbers on your property. Fluazuron products (e.g. Oztik, Tickstar, Tik Shot, Acatak) should be applied strategically, initially at the onset of the tick season and preferably prior to tick activity. Re-treat when ticks are first seen, but not less than six weeks from the previous treatment. The maximum number of treatments per year is three.

Fluazuron is absorbed orally via grooming and through the skin. It should be applied in two 7 cm wide bands, either side of the spine from shoulder to tail. This maximises absorption and ensures effective concentrations of fluazuron. There is no need to treat suckling calves. Calves will receive adequate protection from their mother.

Check the dosage rate as they can range from 3 mL per 50 kg to 5 mL per 50 kg if containing fluazuron. The only product to date that comes in 1 L is Acatak Duostar with a dosage rate of 5 mL per 50 kg.

Protocols for spraying cattle for ticks

If you reside in a tick area and wish to take your cattle to a show or an event, or sell cattle to clients who reside in a tick-free area (referred to as clean country), then

you will need to ensure your cattle are totally free of ticks when inspected at the relevant crossing. If there is even one live tick on one animal the whole load will not be allowed to proceed any further. You can use a spray, a pour-on or an injectable. These are my protocols.

Rules for spraying: it will depend on the season and prevalence of ticks on the stock, but I prefer the animals receive at least three spray treatments with either Taktic EC or Amitik. These treatments are to be done every seven days.

- Cattle are to receive a treatment a minimum of four days and a maximum of seven days prior to being presented to the Department of Primary Industries.
- Make sure a high-pressure sprayer is used, that saturates all the hairs. A hand sprayer is not strong enough to be effective.
- Areas that need special attention are tail heads, brisket, head, dewlap, scrotum, udders, armpits and belly. Do not spray down the ears.
- Make sure the animals are saturated all over and walk away dripping wet.
- Spray from both sides of the crush.
- Calves need to be sprayed as well.

Rules for injectables/pour-ons: cattle are to receive a treatment a minimum of six days and a maximum of nine days prior to being presented to the Department of Primary Industries.

Buffalo fly

Dark-coated cattle, bulls, older cattle and those in poor condition are the most susceptible. The tropical breeds do not appear to be as severely affected as the British and European breeds.

Each fly feeds up to 10–40 times per day by puncturing the skin and taking blood. The discomfort caused to the cattle is obvious as they continually try to shake off the flies, resulting in weight loss, sore and irritated eyes, lesions and pinkeye.

In beef cattle, a moderate infestation of some 200 buffalo flies recorded losses that averaged 15 kg over a 100-day fly season. The constant irritation also causes milk production to reduce, which will have a flow-on effect for the calves and in turn reduce their weight. Dairy cows affected by buffalo fly had milk production reduced by more than 500 mL per day.

Left untreated, buffalo fly can cause nasty lesions that can not only permanently damage the hide but also cause infection and more discomfort. Untreated cattle are more likely to develop pinkeye that those that are treated. Pinkeye can cause blindness in one or both eyes, is highly contagious and requires treatment.

There are a number of treatment options. The method you choose will depend on the number of head you are treating and how often you can monitor your animals.

There are three major chemical groups used in buffalo fly control programs:

- synthetic pyrethroids (SP);
- organophosphates (OP) and carbamates;
- macrocyclic lactones (MLS).

Combination OP and SP sprays are also available. Remember to check the withholding period (WHP) and export slaughter interval (ESI) of any chemicals used, especially if you are planning to sell livestock to the markets. These chemical controls can be applied in various ways.

- Insecticide-impregnated ear tags: most tags are effective for 10–16 weeks. Some brands require two tags and others only one. We use brands that require only one tag due to less handling and fewer holes in the ears, which is especially important for our show team.
- Pour-ons: for small numbers of cattle this is often a very effective method although application will need to be more regular.
- Sprays: usually a cheaper option requiring multiple treatments.
- Back rubbers and rubbing posts: these allow the cattle to self-apply. Success depends on how often cattle use them.

By following a few simple strategies you can reduce the incidence of chemical resistance, which will not only benefit the welfare of your cattle but also our environment. Alternating between the two main chemical groups (OP and SP) is highly recommended. Do not use OPs for more than two seasons in a row or SPs continuously for more than one season. Remove ear tags at the end of their effective life-span, which is normally 10–16 weeks depending on the product – make sure you check.

Non-chemical control options include the following.

- Organic back rubs: back rubs are soaked in an organic solution, which is distributed as the cattle rub themselves against it. Very effective.
- Culling allergic cattle: this is a long-term strategy for prevention of the impact of buffalo fly in a breeding herd. Care must be taken not to confuse the lesions caused by stephanofilaria with allergy, as culling the former will have no impact on breeding a herd with buffalo fly resistance.
- Buffalo flytraps: buffalo flytraps can reduce fly populations by up to 70%. They are mainly used in situations where cattle can be trained to use them on a regular basis such as in dairies or cell-grazing situations.
- Dung beetles: dung beetles reduce buffalo fly populations by removing or spreading dung so that the flies cannot breed in it.

Q. What months do you treat for buffalo fly?

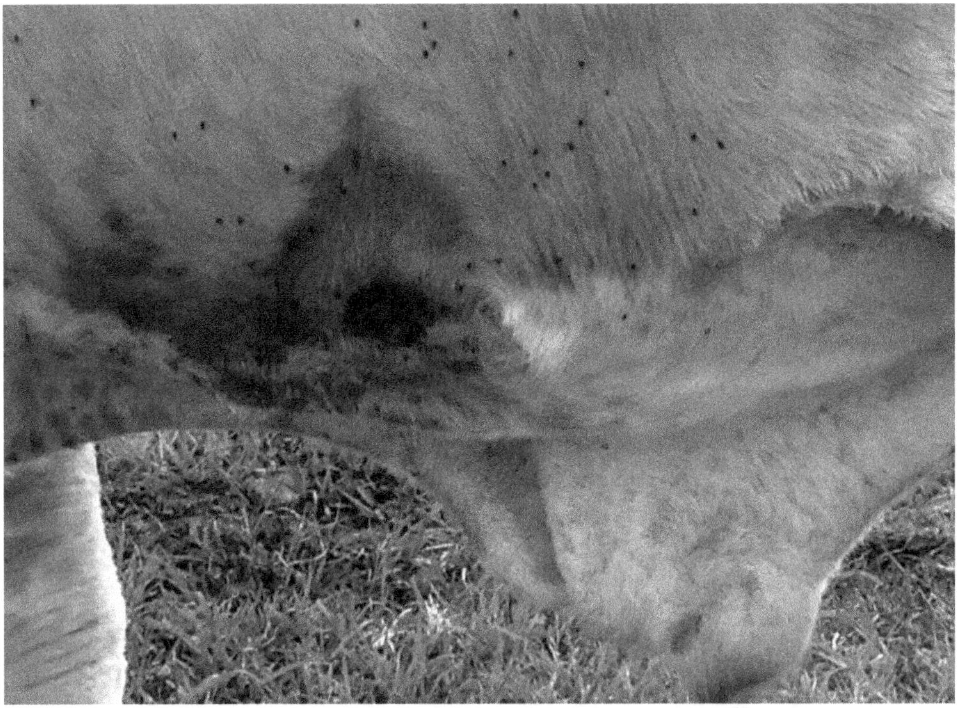

Left untreated, buffalo fly can cause great discomfort and, in severe cases such as this, bleeding.

A. We use ear tags when the flies start to appear, which in my area is late December. We use the brand where only one tag has to be applied as it makes for easier handling and fewer holes in the ears, which is a consideration for show stock. The tags should last three months. Some brands are better lasting than others; it depends on the active ingredient.

Lice

There are two main types of lice, i.e. biting and sucking, and both can affect cattle. They irritate and distress the cattle causing them to rub, scratch and bite. Sometimes they rub so much that they bleed. Lice are most common when times are very dry. They are also common in the cooler months as they prefer the denser winter coats. Fat, healthy cattle rarely seem to be affected by lice.

Lice are spread by direct contact with other animals and not infectious to any other species of farm animals.

The most important point to remember is that most treatments will not kill the eggs on the body, so follow-up treatments two to three weeks later are required for effective control. Prior to treatment, identify whether the lice are the biting or sucking species as not all products kill both. There are three main methods of treatment application:

- hand sprays, which are easy and effective for small numbers;
- pour-ons;
- injectables.

Q. How do you know if your cattle have lice? What are the signs?

A. Cattle will rub against trees and fence posts, often resulting in bare patches of skin. Young, thin and poor animals are most susceptible. Cattle that reside in areas of relatively high humidity very rarely become affected by lice.

Tip: Remember when applying pour-ons and drenches to check the withholding period (WHP) and export slaughter intervals (ESI). The WHP is the period following treatment when cattle are not allowed to be processed for domestic consumption. The ESI is the period following treatment when cattle are unsuitable for export processing. Failure to comply could incur significant fines if residues are detected in an animal at an abattoir.

Internal parasites

Internal parasites include worms and liver fluke. Good internal parasite control programs offer a highly cost-effective and sustainable means of increasing production.

Worms

Moisture is essential for worm survival. Round worms (nematodes) and flat worms (platyhelminthes) are the main two types of worms you will encounter. The major action of nematodes is in the stomach or small intestine where they attack the gut wall, consuming blood, causing anaemia and reducing the nutrient absorptive capacity of the tissues, resulting in further debilitation through nutrient deprivation. The life-cycle of a parasite is approximately three weeks, so very high populations can build up over three or four breeding cycles.

Left unchecked, they can cause serious and dangerous effects on an animal's performance. Young cattle, older cattle and working bulls are particularly susceptible and can suffer from loss of condition, anaemia, bottle jaw, dehydration, scours, rough coat and, in severe infestations, death. Calf deaths can occur 10–12 weeks after initial infestation, especially if animals are stressed or underfed. Immunity develops with age and exposure, taking between one and two years to fully develop.

Good internal parasite control programs offer a highly cost-effective and sustainable means of increasing production. There are a number of ways to control

these parasites, by the far the easiest for small numbers is via pour-ons. Most of the pour-ons are broad-spectrum and control a whole range of gastrointestinal worms as well as lice, mites and ticks. Liver fluke is quite specific to certain areas so check with your vet, the Department of Primary Industries or neighbours to see if your area is prone. It is a good idea to ensure that all products used are dung beetle friendly.

Methods of application can include oral and injectable drenches:

- pour-on drenches (i.e. ivermectin- and moxidectin-based drenches);
- injection products (i.e. ivermectin, moxidectin and abamectin);
- oral drenches (i.e. levamisole);
- vaccines.

We treat all our calves every three months after birth to one year of age. We do not drench our senior cows unless there seems to be a specific problem affecting them, such as poor nutrition due to drought or old age. Remember that older animals are just as susceptible as younger ones but are often overlooked in worming programs.

A range of strategies can reduce the use of chemicals, including pasture management and good nutrition. In the long term, these prevent chemical resistance. Properties that are understocked seldom need to drench for worms. Re-infestation can be prevented by regularly resting your paddocks, which helps prevent the emerging larvae being eaten by grazing cattle. Once cattle have been wormed, move them to a clean paddock to reduce the chance of re-infestation. Changing the active ingredient also helps avoid resistance build-up. Drench resistance is on the rise in Australia and recent research findings published in *Veterinary Parasitology* confirm that it is essential to use combination drenches as part of an integrated package of resistance management strategies.[3] The website of Animal Health Direct provides a list of chemicals and their active ingredients, which is valuable to print.

Worm egg counting is one of the most useful management tools for a producer. Faecal egg counts can be undertaken by most veterinary laboratories and private veterinarians. Private companies such as stockwatchlab.com.au and wormboss.com.au offer an excellent service. Another website (wormtrax.com.au) is excellent at showing what worms are most common in your area. Just type in the postcode. Be careful, though: this website is owned by Zoetis and the recommendation for treatment (if any is required) is always to use Dectomax, a Zoetis product.

Q. What is the best month to drench for worms?

A. We drench all calves every three months from weaning, i.e. in their first year they get four drenches. We do not drench our senior cows and bulls at all unless there seems to be a specific problem. We change the active ingredient used in our product to prevent resistance build-up and regularly rest our paddocks to help break the life-cycle of the parasites.

Q. What products do you use for worms?

A. I like to use either pour-ons or injectables. It also depends on the time of the year as to whether I want a product to control only worms or to control ticks as well. Some of the injectables include:

- Cydectin Long Acting: this has the longest persistent activity of any product that I know of, lasting up to 120 days. It cannot be used on cattle under 100 kg or over 500 kg;
- Dectomax Injectable: has the advantage that it is suitable for smaller and heavier animals.

Some of the pour-ons include:

- Genesis;
- Wintix;
- Ausmectin Pour-on;
- Acatak Duostar;
- Tickstar.

Dehorning

The *Veterinary Surgeons Act 1986* defines dehorning as 'the removal of the horn of an animal by methods which destroy or remove the keratin-producing cells and structures at the base of the horn'. The Australian Model Code of Practice for the Welfare of Animals (cattle) states that it should be done as young as possible, i.e. within three months of birth, to reduce stress and minimise animal welfare concerns. Under state legislation dehorning of stock over 12 months of age is illegal without the use of anaesthetics, while the Model Code of Practice recommends under six months of age.

Dehorning is painful and can be completely avoided by breeding polled animals. Horns need to be removed as they:

- can hurt and damage other cattle;
- pose more of a risk to handlers in the yards;
- are illegal in show cattle;
- do serious damage to hides and carcasses;
- are harder to handle when applying NLIS tags, ear tags, tattoos etc.

Cattle should be dehorned on dry cool days to allow the wound to dry quickly with the minimum risk of infection. Never dehorn cattle in wet weather, because the healing rate is decreased and the risk of infection increased. The best time to

dehorn is before the horn bud attaches to the skull, as this results in less bleeding. The horn bud generally attaches at around two months. Dehorning is best done as soon as the horn bud is visible as this causes animals the least amount of discomfort.

There are a few different methods.

- **Dehorners**: cup and scoop dehorners are usually used when the animal is slightly older. The horns are cut from the animal's head. Having spoken with a number of Dexter breeders who have to deal with horns, the majority seem to use this method.
- **Dehorning knife**: used on animals two to three months of age.
- **Electric iron debudders**: this method is used on very young cattle. A hot iron is placed over the horn buds and burns them. It is very effective.
- **Caustic pen:** can be used on calves from one to four weeks. The acid very effectively burns the horn buds. Some literature advises against this method as the chemicals can spread into the eyes if the skin gets wet.

This cow needs to have her horns cut as they are about to grow into the side of her head.

Castration

Male animals that will not be required for breeding need to be castrated, after which they are called steers. Castration is the process of removing the male testicles. It is also referred to as marking.

Castration is done for a number of reasons, as steers:

- fatten more easily;
- are less aggressive;
- are easier and safer to handle;
- make better beef than bulls.

The Australian Model Code of Practice for the Welfare of Animals (cattle) states that castration without anaesthesia should be done prior to six months. Castration of animals older than six months is illegal under some state legislation unless undertaken by a vet.

We castrate six weeks before we wean to help reduce the stress and prevent the calves losing too much condition, i.e. usually at about four to five months of age. If left uncastrated for an extended time the calves will have an excess of male hormones, making them too 'bully'. If you castrate them too young then you're removing the natural testosterone that helps them grow, and this reduces the growth rate of that calf. Testosterone promotes muscle growth.

Good hygiene is essential and it's vital to keep stress levels to a minimum:

- ensure that cows and calves are separated from each other for as short as time as possible;
- perform the procedure in the coolest part of the day;
- return the animals to clean, shaded grassed areas afterwards;
- ensure they have adequate water;
- avoid castrating in wet, humid and dusty conditions.

There are a number of methods. They are categorised into immediate (surgical, using a blade) or delayed (non-surgical, using an Elastrator ring or Burdizzo emasculator). We use both an elastrator and surgery. Surgery is used on steers that will be shown as led steers as judges like to see what's called the cod – the purse where the testicles were. Elastrators do not leave this. The cod is one of the areas that steer judges observe for fat deposits.

- **Elastrator:** a simple device that places a rubber band over the testicles and cuts off the circulation. The advantage of this method is that it is easy to do and there is no blood. Make sure both testicles have descended and have been pulled through the band. The pressure cause the testicles to wither and die but beware, tetanus infections are higher with this method. Under the cruelty code this method must be performed on calves before two months of age. The

Australian Model Code of Practice for the Welfare of Animals (cattle) recommends this method is for calves up to two weeks of age.

- **Surgical**: this is the process of cutting the testicles out of their purse with a scalpel or very sharp knife. It is quick and effective but you will need an experienced person to show you how to perform the procedure. Animals can bleed a lot and may get infections or flystrike, so avoid doing surgery in very hot weather and especially not in the middle of the day. Clean sterilised instruments must be used at all times, with the antiseptic solution and scalpel blade changed every 15–20 calves. If infections occur, a shot of penicillin is very effective in controlling them.
- **Burdizzos**: these are used to clamp and crush the cords supplying blood to the testicles, causing them to wither. This method involves no bleeding. Without a blood supply the testicles wither inside the scrotum and disappear over four to six weeks. There is little risk of infection due to the lack of open wounds.

Q. What is the best time to castrate? We were advised to do it when we wean.

A. I like to castrate at least one month before I wean to reduce the stress level to calves. The calves will be in pain whichever method you use, so returning to their mothers to suckle will comfort them and they will be less likely to lose much weight. At the same time I make sure the calves have had their first injection of 7-in-1.

Meat and Livestock Australia has produced an excellent book: *A Guide to Best Practice Husbandry in Beef Cattle Branding, Castrating and Dehorning* (2007). The NSW Department of Primary Industries website also has an excellent fact sheet.

Nose rings

Nose rings are used in the control of show cattle: the rule is that all bulls over 12 months of age must have one. They are also used on led steers and some breeders use them on unruly females.

A hole is punched through the connective tissue of the nose and a ring is placed through the nose. The ring is secured with a screw and remains there until removed. Rings can be made out of brass, aluminium and copper.

You can use anaesthesia and a vet to nose ring your cattle but most stud breeders, including myself, do it themselves very effectively with a nose punch. It's a quick and effective method similar to getting your ears pierced. If a bull won't let anyone near its head or nose this is often based on a past painful experience.

If an animal has finished his showing days and you plan to turn him out in the paddock, take his nose ring out. It can be dangerous if it gets caught on a fence. If this happens the animal will stand there for days rather than have the pain associated with pulling back on the ring.

Other possible health problems

There is a lovely saying in medicine, 'Common things occur most commonly.' The same applies to cattle so I have mentioned only problems that we see on a regular basis. Of course there will be times when an animal will contract a rare illness, which on most occasions will require veterinary assistance.

I am a great believer in early intervention rather than waiting to see what happens. A shot of penicillin on day one or Opticlox eye ointment at the first sign of a weeping eye can often save days of further treatment. This is because treatments often take 48 hours to take full effect: if you wait until a calf is really ill then it could get a whole lot worse before it even starts to improve – and that could be too late.

Grass tetany

Grass tetany is a nervous disorder caused by a magnesium deficiency in the blood. It occurs mainly in New South Wales, Victoria and Tasmania and is more common in the temperate areas than in the tropical. It is seen in cold, showery and windy weather with cattle on heavily fertilised pastures and grass in the rapidly growing phase. It normally occurs when there is a flush of green feed. Older cows are at greatest risk, usually in winter. Consult your neighbours or the Department of Primary Industries to see if your property is prone to grass tetany. There are a number of preventive measures including:

- feeding hay to increase roughage in the rumen, which will increase absorption of magnesium by the stomach;
- salt licks, as salt will increase the absorption of magnesium;
- magnesium lick blocks;
- feeding Causmag is an effective form of prevention;
- Mag C in the water troughs.

Bloat

This is a serious problem for southern producers in spring. Bloat is an accumulation of excess gas in the rumen, which usually results from cattle grazing lush pastures with high legume content such as clover, lucerne and medics. Death rates can be as high as 20% in high-risk pastures so prevention is better than cure.

Symptoms

- Swelling and distension of the animal's stomach, in particular on the left-hand side.
- Obvious distress – panting, open mouth, lethargy, unwillingness to move.
- No desire to eat.

Treatment

- Slow-release anti-bloat capsules.
- Anti-bloat lick blocks.
- Water medication such as Teric Bloat Liquid.
- Trocar (boat knife).

Prevention

- Introduce cattle slowly to high-risk pastures, especially hungry or poor cattle.
- Make sure they have adequate roughage (feeding dry quality hay is useful).
- Try to avoid having too many legumes in your pasture and increase the grass content.

Freemartin

This occurs when twins of opposite genders are born. The female may be infertile, as the male hormone can prevent the female developing female characteristics. This is why twinning is not encouraged in cattle. If twins are of the same sex, there are no known associated problems. However, I have had a number of sets of twins of mixed sexes and none of the females have been infertile but it is worthwhile being aware that it can occur.

Navel infections

Small breeds seem to be more prone to this infection, perhaps due to the calves' close proximity to the ground. Calves are normally affected in the first few days: infection is caused by bacteria or dirt and dust entering via the umbilical cord at or soon after birth. The area will be swollen. If left untreated, this swelling can result in the formation of a pus-filled abscess at the site. The calf may have a fever and reduced appetite. Untreated, the problem can lead to sepsis and death. Treatment consists of a course of antibiotics. I use long-acting penicillin.

Retained afterbirth

The placenta is normally expelled two to eight hours after calving. Occasionally a cow can be seen with a retained afterbirth and, although it may seem unsightly, it normally will fall out by itself. If this does not occur within 12 hours then put on sterile gloves and give it a gentle pull. An injection of oxytocin will help stimulate the muscles to contract and expel the afterbirth. If after 48 hours there is still considerable afterbirth showing, then you will need a vet to assist. In extremely hot weather a retained afterbirth can result in flystrike and serious infection, especially if it is dragging on the ground.

Ringworm

Ringworm is caused by fungi and moulds. It is contagious and should be treated with iodine solution. It will not cause the animal any harm but there will be excessive rubbing due to itchiness. Some shows will not allow animals to be exhibited if they are infected with ringworm.

Scours (diarrhoea)

The most common cause of scours is a change in feed. Cattle that are suddenly put onto lush green feed or fed prime lucerne hay will nearly always scour. Feeding some dry hay can help, but there should be no need to be concerned as long as the animals seem healthy.

A high infestation of parasites can also cause scours and this will need to be treated.

Calves are often affected by scours and they need to be monitored closely as affected animals can dehydrate and die. They are more likely to be affected in the first six weeks and if they are calves from heifers or from cows that are in poor condition. Severe copper or selenium deficiencies can also be responsible for scours. You will observe sticky white or yellow diarrhoea around the tail, and blood is sometimes present. There are a number of very effective treatments:

- scour tablets, which are the easiest to administer and work very effectively;
- a number of oral soluble powder treatments;
- pastes – we use D-Scour paste which is excellent;
- if there is severe dehydration you will need to administer electrolytes.

Show cattle are also susceptible to scours due to overwatering and change of feed.

Warts

Warts are caused by a papilloma virus that usually affects only younger cattle between 12 months and two years of age. There are a number of different types of warts, ranging in size from a pea to a tennis ball. The most common place for them to occur is the head, neck and shoulders but they can spread to other parts of the body. Although they are unsightly, they usually cause no permanent damage. Some types of warts may involve the venereal regions where they can cause pain, disfigurement, infection of the penis of young bulls, and dystocia when the vaginal mucosa of heifers is affected.

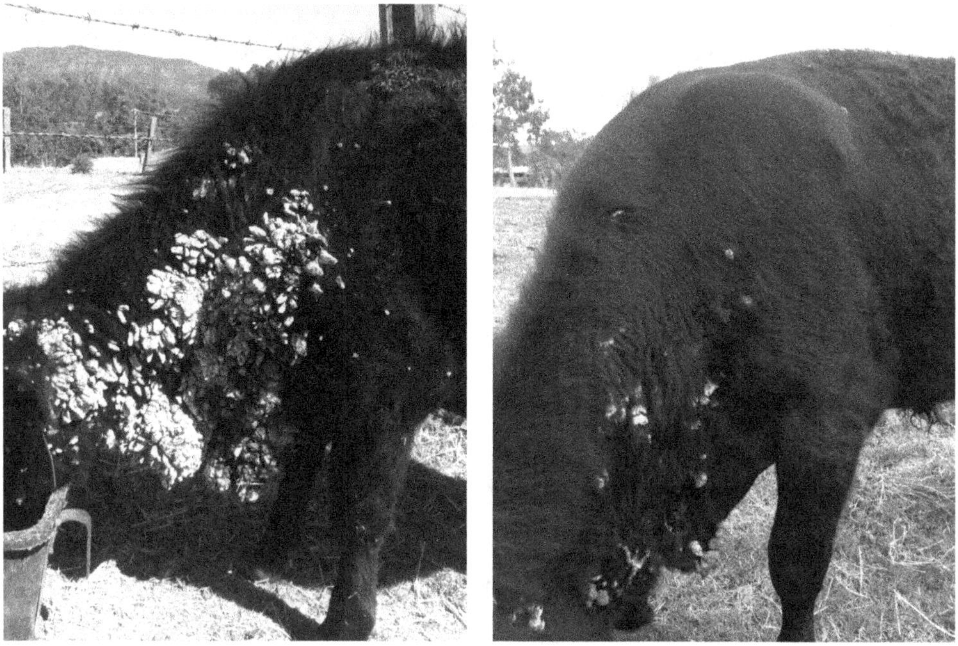

Left: An animal heavily infested with warts. This was the worst case I have ever seen but all the warts disappeared within six months. Right: This is the same animal four months after the outbreak.

If large parts of its body are covered in warts then an animal may not thrive. Sometimes warts that bleed may become flyblown; these will need to be treated.

The incubation period, i.e. the time a wart takes to develop after infection, is normally three to eight weeks but may be longer. Warts are very contagious and the virus is transferred by cattle rubbing on gateposts, fence posts, trees or each other. They can also be transferred through equipment such as tattoo guns and crushes. The virus gains entry through skin abrasions. Unfortunately, once you have introduced warts into your herd it is very hard to get rid of them. An entire group may be exposed but not all will contract it.

Affected animals may be isolated from susceptible ones to reduce the chance of spread but, with the long incubation period, many are likely to have been exposed before the problem is recognised. Mustering, yarding, dipping and other husbandry practices should be kept to a minimum during any noticeable outbreak of warts, to avoid stress.

There is no known vaccine for warts. As stated, one method of treatment is to isolate infected cattle. Another is to cut one of the warts so that it bleeds, as this sometimes results in immediate regression. It is believed that this method produces an active immunisation that causes all warts to fall off. Disrupting the wart encourages the animal's immune system to create antibodies; the virus in the disrupted tissue comes into contact with the bloodstream if the area bleeds a little.

There are lots of other old-fashioned remedies, which include covering the wart with Vaseline and banding the wart so it eventually falls off. Alternatively, the best treatment is time ... if you simply leave them they will disappear after six months, never to reappear on that animal. Once an animal has had an infection of warts it will never get them again. Some show regulations do not allow heavily infested cattle to be shown, so you will need to check this at the time of entry.

Dermatophilosis (rain scald)

Caused by the bacteria *Dermatophilus congolensis*, this problem is commonly associated with wet weather. I first encountered it in 2014 in one animal, and in 2015 a severe outbreak occurred when we had 30 cm of rain in one week. Clinical signs developed at a rapid speed. At first I thought it was ringworm, as it looks similar.

Moist circular patches, often with a crusty scab, appear. Textbooks say the symptoms are typically seen on the muzzle and tips of ears and spread rapidly. My cattle showed lesions all over the body.

There is no completely effective treatment, particularly for severe cases. Many producers often do not treat mild cases. Most animals will recover spontaneously within three weeks of the initial infection. Animals with lesions over more than 50% of their body should be treated as they can become seriously ill. Isolate clinically affected animals to help reduce the spread and break the infective cycle. Checking zinc levels in the feed of cattle is also recommended, because outbreaks have been associated with zinc deficiencies.

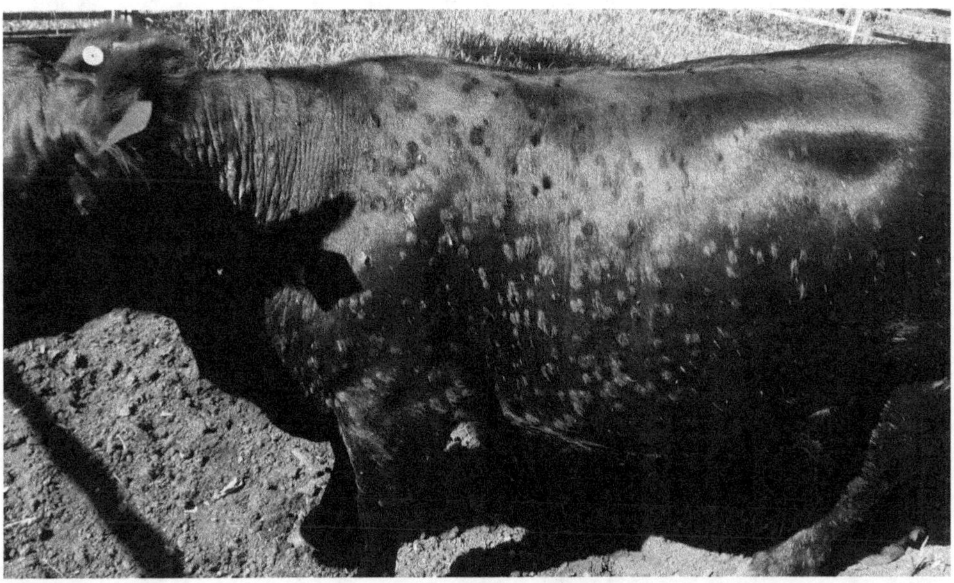

The rain scald on this animal cleared up with one treatment of Condy's Crystals, which I got from the pharmacy.

Dr Sandi Jephcott gave me the following advice, based on her experience in Papua New Guinea. Her method of treatment involves Condy's Crystals (potassium permanganate). Mix the granules with water until you get a strong colour, generally a tablespoon of crystals to a 10 L bucket. Put the cattle in the crush and rub or brush off the crusted, matted hair then get a sprayer/backpack full of the Condy's Crystal mixture. Spray over all the area that had the matted hair. Remember to disinfect any brushes you use. Dermatophilosis can be transmitted to humans so use gloves and wash with antibacterial soap. Follow up initial treatment with long-acting alamycin. Other recommended treatments include penicillin.

Unusual deaths

There is a saying that is never of any comfort when you lose an animal but it is true: 'When you have livestock you will have dead stock'. It is part of life and, while it can be upsetting at the time, trying to prevent it happening it again is the best action to take. I have lost animals to some very unusual incidents, most of which were preventable:

- lightning strike;
- brown snake hibernating in a round bale of hay;
- choking in a crush;
- plant poisoning (green cestrum);
- a calf drowning in a molasses tub;
- a calf drowning in an old well;
- an animal's head stuck in the fork of a tree.

General tips

Transporting cattle

Any animals that we take on long journeys are given some probiotic paste or gel just before they are loaded. This prevents dehydration and definitely makes a difference. We also use it on our show animals when they are housed for long periods at royal shows, or for cattle that have an upset tummy.

- Don't load mixed mobs of unfamiliar stock. They will need at least two days together for familiarisation before they can travel.
- Lock cattle in the yards with access to water but no green feed the night before they are due to travel. This is called 'emptying out' and prevents them from having to stand in a truck full of sloppy manure which can cause them to slip and make the journey more uncomfortable, particularly if there are young calves.
- Load cattle quietly. It is better if they are loaded tightly to give each other support.

- Don't move stock during very hot weather or in the heat of the day with no shade. Move them at night when it is cooler. Likewise, don't move stock in the middle of winter in unprotected vehicles.
- If you are travelling a long distance, such as interstate, rest stops often make the journey more stressful with loading and unloading. It's much better to go straight to your destination, within reason, then give the stock at least 12 hours rest. Of course this will depend on how well the stock are travelling and on their age. If we are driving from Brisbane to Sydney we drive straight through but if we are going to Victoria then we stop overnight, allowing the cattle to rest and feed.

Buying cattle from other regions

Tip: Make sure that the floor of your transport offers the cattle a good grip. I can't tell you the number of times people have arrived to collect cattle in a small box trailer with a plain steel floor. Once such a floor is covered with manure and urine it is extremely dangerous and slippery, and animals can do themselves serious and permanent damage. Mesh makes the best flooring for these situations, even old carpet works well as long as it is secured to the floor. For long distances in a covered truck, a floor covered with thick straw, sawdust or open-weave rubber mats works very well.

Cattle from other regions will take some time to adjust to a new environment. My experience is that it takes at least 12 months for cattle to adjust, especially if they have come from interstate. For the first three months they are usually unaffected then they seem to lose condition for quite some time before they finally adjust to the new climate.

Quarantine

It is a good idea to separate any newly purchased cattle for a month. This will allow you to vaccinate and worm them and make sure they are not carrying any contagious diseases.

Suggested management plans

We have an autumn and spring calving and these management plans are based on that (Tables 8.3, 8.4). Some producers on smaller acreages may prefer to run a bull with the herd all year, which is perfectly acceptable. The reason we have set calving times is for ease of management so that all calves can be weaned, vaccinated and marked at the same time.

Table 8.3: Annual management plan for autumn-calving herd

Autumn	**March** Calving Worm last year's calves 1st dose of Pestiguard for last year's heifers	**April** Calving Vibromax bulls prior to joining	**May** Calving 2nd dose of Pestiguard for last year's heifers Vibromax bulls prior to joining
Winter	**June** Join cows Worm calves 3 months of age	**July** Join cows	**August** Join cows Tick fever all calves 1st dose of 7-in-1 for current calves Mark (castrate) calves
Spring	**September** Worm calves 2nd dose of 7-in-1 for current calves Booster 7-in-1, BEF and Pestiguard for all cows	**October** Wean calves Give pinkeye vaccine	**November**
Summer	**December** Worm calves	**January** Treat for buffalo fly	**February**

Table 8.4: Annual management plan for spring-calving herd

Spring	**September** Calving Worm last year's calves 1st dose of Pestiguard for last year's heifers Booster 7-in-1, BEF and Pestiguard for all cows	**October** Calving Vibromax bulls prior to joining Give pinkeye vaccine	**November** Calving 2nd dose of Pestiguard for last year's heifers
Summer	**December** Join cows Worm calves 3 months of age	**January** Join cows Treat for buffalo fly	**February** Join cows Mark calves but not in excessively hot weather Tick fever all calves 1st dose of 7-in-1 for current calves
Autumn	**March** Worm calves 2nd dose of 7-in-1 for current calves Booster 7-in-1 for cows	**April** Wean calves	**May**
Winter	**June** Worm calves	**July**	**August**

9

Methods of identification

Management tags

Most owners used some method of ear tag to identify their animals, and tags are recommended. We use a colour-coded system, i.e. red tags for purebreds, blue tags for 1st cross, yellow tags for 2nd cross and so on. We try not to tag very young calves as it weighs their ears down and sometimes causes permanent damage. Tags can be preprinted with a range of information such as stud name, stud prefix etc. Our tags are preprinted with our stud name and we use a pen to write on them each animal's unique identification number and date of birth. On the back we identify the dam. One disadvantage of ear tags is that they can be lost; for example they can get ripped out on fences and posts, leaving an unsightly looking ear.

Tags come in a variety of shapes and sizes. Some are one-piece tags (e.g. Z Tags) and others are two-piece, consisting of a male button and a female tag. It is possible to have a visual tag that matches your NLIS device, i.e. your management number marked on the NLIS tag is also printed on the visual tag. Some companies can also integrate DNA technology, for complete traceability.

Tip: If handwriting information on a tag, make sure you use a special permanent tag pen and overwrite the information a couple of times to prevent fading.

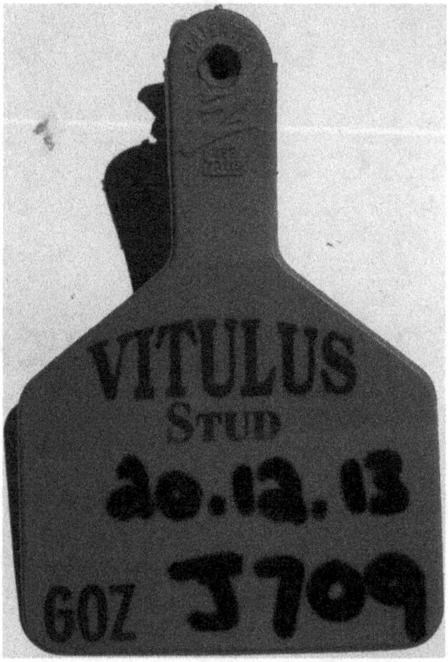

Our management tags are always placed in the left ear so they are easy to read when cattle come into the crush.

Place the tag in the centre of the ear, avoiding veins. If you have to tattoo your animal do so first, so the tag does not obscure the tattoo. You will need to buy an applicator to suit your brand of tags. Every brand of tags has a required applicator so you are probably better off finding a brand you like and sticking to it, as applicators are expensive.

Tattooing

This is a permanent system of identification and is mandatory for the majority of stud stock and animals exhibited at royal shows. Done correctly, a tattoo should last the lifetime of the animal. Usually one ear will contain the stud prefix and the other the animal's unique number, which is usually prefixed by a letter of the alphabet.

The tattooing numbers are made up of a series of very fine sharp spikes. They fit into pliers which are squeezed together tightly in the ear. Ink is then rubbed into the indentations made by the pliers, leaving a permanent tattoo.

There are a few tips to tattooing correctly.

- Clean the part of the ear to be tattooed thoroughly, with a rag dipped in methylated spirits.
- Apply some paste on the area to be tattooed. Make sure this is not on a vein and is in the centre of the ear.

- When placing the numbers or digits in the pliers, place them from left to right when looking at the pliers.
- Check them on a piece of paper first to make sure you have the right order. This is very important!
- Apply the pliers with a very firm grip. You want to hear the ear crunch and possibly some blood will appear.
- Apply the paste again thickly, making sure it fills the indents made by the pliers. A toothbrush is good for this.
- Green paste is the best to use on animals with dark ears.
- Avoid tattooing tiny calves as the ear may be too small for it to fit and the tattoo may grow out with the ear.
- Beware: tattooing is a messy business and you can end up with tattoo ink everywhere.

Tattoos can usually be read only with the aid of a torch and when the animal's head is restrained. You often have to clean the ears first to see the numbers properly.

Q. How do we know what letter of the alphabet to use?

A. All stud animals are allocated a unique identification number, by their breed association. Each year, nationally across all breeds, the same letter of the alphabet is used to prefix the animals that are registered in that particular year; for example in 2015 the letter is L. All animals born in 2015 at our stud will have their stud names prefixed with the letter L. This is not compulsory but it is common practice in stud cattle. Two letters are not used by breed associations: I and O.

National Livestock Identification System

The National Livestock Identification System (NLIS) is the nationally adopted standard for the traceability of cattle and is enforced by state-based legislation. It is a tagging ID system that offers lifetime traceability of cattle, sheep, pigs and goats. Individual animals can be tracked from property of birth to slaughter, for food safety, product integrity and market access purposes.

NLIS uses machine-readable Radio Frequency Identification (RFID) devices to identify cattle with a unique, unalterable 16-character 'licence plate'. The details about each animal's history, such as changes of ownership and property, as well as information that may affect their suitability for human consumption, are stored in the secure central NLIS database managed on behalf of the industry by Meat & Livestock Australia.

NLIS-approved devices come in the form of an ear tag or rumen bolus/ear tag combination. Each ear tag consists of two unique numbers and your property identification code (PIC) will be included in your NLIS tag. The RFID numbers are inside the tag and are uploaded to the NLIS database with a reader, along with the device's corresponding NLIS ID number, which is printed on the outside of every device and can be read visually. Cattle are tagged with NLIS devices only once in their life unless the device becomes detached. There are a number of accredited devices and several tag manufacturers, so shop around. Some can be integrated with your management tags.

There are rules in the NLIS system.

- Cattle producers must apply a white 'breeder' NLIS device (either an ear tag or a rumen bolus/ear tag combination) to cattle bred on their property.
- Orange 'post-breeder' devices are applied to cattle no longer on their property of birth and not already identified with an NLIS device. This includes cattle that have lost their breeder device after leaving their property of birth.
- When cattle move from one PIC to another, their devices are scanned electronically with a tag reader or the NLIS ID number is read visually and the number is noted. The consignment's movement details are then recorded on the NLIS database and automatic notifications of the movement are sent to the appropriate account holders and authorities via email.
- If cattle are bought or sold through a saleyard or sold to an abattoir, the saleyard or abattoir must record the movement.
- If cattle are bought or sold privately, the person who receives the cattle is responsible for notifying the database of the movement.
- If cattle move between properties with a different PIC, the movements must be recorded on the database even if the properties have the same owner.
- If cattle move to an agistment property owned by someone else, the movement off the owner's property and onto the agistment property must be recorded on the database.

Important: You will need to contact the Department of Primary Industries to obtain an application form to order NLIS tags. It normally has an expiry date of up to two years. Take the completed form to a rural produce supplier or order the tags online. Remember you will need to purchase an NLIS applicator for the tags as well.

Your tag will contain the following information with a 16-character number:

- QILL0052 – PIC (property identification code);
- X-NLIS code to identify the manufacturer;
- B-NLIS code to indicate device type (B = breeder, E = post-breeder);
- G-Breedplan year code. 'O' and 'I' are not used (see Table 9.1);
- 00001 – your unique management number. You can choose to have these numbers match your management tags and can specify the numbers.

Table 9.1: Breedplan code for year of manufacture

Year	Code
2011	G
2014	K
2015	L
2016	M
2017	N
2018	P
2019	Q
2020	R

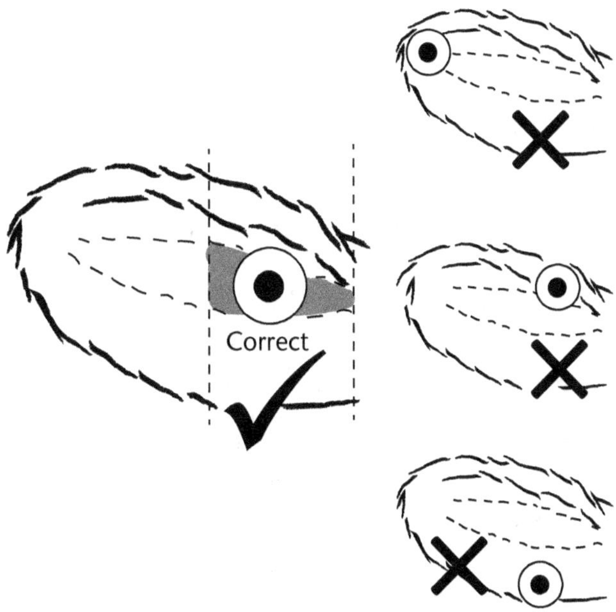

Correct placement of NLIS tags.

Branding

Branding is the placement of a permanent identifying mark on the hide of an animal by destroying the hair follicles and altering hair regrowth. From a welfare perspective, branding is not the preferred method of identifying livestock. Other methods, in particular NLIS devices such as an ear tag or rumen bolus, are preferred. Hide damage at the branding site also decreases the value of a tanned hide.

The requirement to brand cattle varies around Australia (see Table 9.2). Hopefully sense will prevail and the requirements may change as part of the development of new Australian Standards and Guidelines for the Welfare of Animals. Cheek branding is illegal in most states.

When branding, use the following tips.

- Good restraint is essential – use a cradle for young calves and a crush for older cattle.
- Never brand wet, weak or emaciated cattle.
- Brand animals when two to six months of age.
- When hot-iron branding, ensure branding irons are well-maintained and heated to the correct temperature (apply the hot irons for a maximum of two to three seconds).
- Do not fire brand after treatment with a pour-on chemical.

There are three types of branding.

- **Fire branding**: The use of brands should be discouraged in the interests of reduced hide damage. They should be as small as possible and positioned close to the centre line of the body. Butt brands should be placed close to the tail head. When carried out, branding is usually conducted with calf marking at about three months of age. Brand temperature should be such that a clear brand will be possible within three or four seconds. Red-hot irons will cause excessive penetration of the hide. Blue-hot irons are ideal. Wet calves should not be branded, as brands cannot be kept at a suitable temperature and scalding may occur.
- **Freeze branding**: Prior to freeze branding, the animal should have reached its mature size. The area to be branded should be flat. The skin should be clipped and swabbed with methylated spirits. Brands are cooled in liquid nitrogen or a methylated spirits and dry ice mixture, and applied to the clipped area. The duration of application varies according to whether a white hair brand or a bare skin brand is required, from 30–60 seconds.
- **Electric branding**: This would seem to be the simplest method to use. The brand is designed to clip onto a 12 V car battery or mains power, which will heat the branding iron to temperature in 90 seconds and maintain a constant heat, producing an excellent brand. An open fire is not required and this is a very safe method.

Q. Let's say I have purchased cattle from you, and they have arrived on my property. Do I have to tag them or brand them and if so, how and where do I get the tags from?

A. It is law that the cattle already have an NLIS device fitted prior to being moved onto your property and you will need to notify the NLIS database of their NLIS number. This can be read from the device fitted in their ear.

Table 9.2: State branding requirements

State	Requirement	Where to brand
Australian Capital Territory	Not compulsory	
New South Wales	Not compulsory	Area to be branded should be a flat piece of skin
Northern Territory	Compulsory if cattle are to be sold or moved off property. Not required for animals under 8 months. 3-letter brand system where one letter must be the letter T and a symbol brand system	1st position must be registered with brand. Sites include shoulder (near or off), rump, hip or thigh (near or off)
Queensland	All cattle over 100 kg must be branded before being sold. Approved stud cattle sales are exempt, as are cattle going straight from the property to abattoir. Brand must be registered	In Queensland it is illegal to brand on the ribs or cheek. Preferred sites are neck, rump, shoulder, thigh, twist
South Australia	Not compulsory	1st position, off rump and hip 2nd position, near rump and hip 3rd position, near shoulder 4th position, off shoulder
Tasmania	Not compulsory	Not site-specific
Victoria	Not encouraged and there is no official brands register	Not site-specific
Western Australia	Compulsory with specific regional requirements	Fire or freeze branded (two letters and one number on the left shoulder)

A. There are no legal requirements for you to tag or brand animals once they arrive on your property.

Q. I know with horses you obtain their breeding details through their brands but I am not sure how to go about this with cattle.

A. Cattle brands show who owned the cattle at the time the brand was placed on them. They do not provide breeding details.

10

Breeding systems

The aim of every producer is for a cow to have a calf every 12 months. This can be achieved either by natural mating or by using artificial insemination (AI).

Natural joining

A natural joining is by far the most efficient method, as the bull knows exactly the right time. There are a number of options.

- The bull runs with the whole herd all the time, but you must remember to remove heifers before they start cycling. There is no reason why this can't be done, if you want to have calves born throughout the year.
- Set joining times so you can control when your cows calve. The reasons for doing that may be seasonal (you may want to calve down when feed is most abundant) or for annual bull sales (so that calves reach a particular age for a certain month). Some producers like to have a very tight calving season of only eight weeks so therefore the bull is only joined with the females for eight weeks. Our joining is for 12 weeks and we have two joinings a year, so that we have an autumn and a spring calving. Set joining times have some benefits from a management perspective in that all our calves can be weaned together and vaccinated together, and the heifers will all be the same age when they are of age to be joined.

Artificial insemination

AI is the physical placement of semen into the reproductive tract of females with the aim of achieving pregnancy. AI is far more common in the dairy industry than it is in the beef industry. We use AI on a regular basis to bring in different genetics, and genetics that are no longer available by live matings. It saves us having to run so many bulls and at the same time broadens our genetic base.

A number of small acreage holders often assume that AI is a cheaper option than having to purchase a bull. Be warned: this is not always the case. For an AI program to be successful you need to have the time to observe your cows (often twice a day to make sure they are on heat), you need to have a facility to store the semen close by so that you can access it when you need it and you need to make sure your AI technician is available if and when you need them. Not all cows cycle on the same day at the same time. Some may be 24 hours apart, so you need to make sure your technician will be available. Also, even the best AI rates are only about 80%. If a cow doesn't conceive you will have to have to run another program in four weeks' time and even then there is no guarantee it will work for every cow. Successful AI is all about timing.

Advantages:

- the ability to use sires of superior genetic merit;
- improving production traits in your cattle operation;
- the ability to mate specific sires to individual cows;
- reducing the number of herd bulls needed in your cattle operation;
- broadening your gene pool;
- can produce a very tight (shorter) calving season.

Disadvantages:

- cattle need to be in good condition and on a rising plane of nutrition;
- cost of consumables required for the program;
- cost of semen and storage equipment;
- time-consuming, as animals need to be observed regularly;
- synchronised programs are time-consuming with daily injections;
- requires an experienced AI technician.

Conducting a program

There are a number of ways to conduct a program and you will need to liaise with your technician to see which method suits them the best. We do our own AI, using the fixed-time AI method. We also use CIDRs, which are devices that contain the natural hormone progesterone. These are placed intravaginally to release progesterone at a controlled rate into the bloodstream. The most important part of any program is being able to detect when the female is on heat. There are some aids available to help you.

- Heat detectors and tail paint can be used as an aid to determine if a cow is on heat. Detectors are more expensive than tail paint but are easier to read, require less maintenance once applied and result in higher detection rates. There are two types available, either pressure-mounted 'tubes' or scratch-off 'patches'. They are applied, by being stuck, to the rear portion of the backbone of the cow. Cows on heat will stand to allow other cows or bulls to mount them, the detector responds to the pressure by changing colour and so is easily visible.
- Tail paint is sprayed onto the animal and turns to powder when the animal is mounted.

'Teasers' are usually males that are incapable of serving a female (e.g. a steer) but they can be used to detect a female on heat. You need to ensure they are free of venereal diseases.

Insemination on observed heat

Oestrus, or heat, is the period in the reproductive cycle when the cow will allow mating. Spend at least 30 minutes in the morning and 30 minutes in the evening observing your cows for oestrus. Early morning and late afternoon are the best times to observe. The visual signs include cows becoming restless and mounting each other. As a cow gets closer to ovulation she will stand herself (called standing heat) to allow other cows to mount her. A clear stringy mucus will be discharged from her vulva. Her vulva may become swollen and hair may be rubbed off her back from other cows mounting her. As each cow is observed as being on heat, she is AI'd.

Fixed-time AI

FTAI is the newest program and claims an 80% conception rate. The benefits are that cattle have to be brought into the yards only three times, heat checking is not required and labour costs are greatly reduced.

Program

1. **Day 0:** Insert CIDRs and inject with 2 mL Cidirol for cows and 1 mL Cidirol for heifers.
2. **Day 7:** Remove CIDRs and inject with 2 mL Prostaglandin (PG).
3. **Day 10:** AI and inject with 1 mL GnRH. AI needs to be timed at 60–66 hours after CIDR is removed; this is critical. Inject the GnRH at AI.

GnRH is a hormone naturally produced in cows that causes the cow to release another hormone – luteinising hormone (LH). This hormone, in conjunction with follicle-stimulating hormone (FSH), enhances the growth of ovarian follicles that contain the developing egg. Large amounts of LH cause ovulation. After ovulation, a corpus luteum (CL) forms on the ovary and produces progesterone, which prepares the uterus for pregnancy and prevents the female's return to heat.

Synchronisation with insemination on observed heat

Oestrus synchronisation involves all cows so that most will display oestrus and ovulate together. We do this using progesterones which block the oestrous activity. They are administered by intravaginal implants and remain *in situ* on average nine to 10 days depending on the brand. On their removal the oestrus cycle recommences in a synchronised manner.

Program

1. **Day 0:** Insert CIDRs and inject with 2 mL Cidirol for cows and 1 mL Cidirol for heifers.
2. **Day 8:** Remove CIDRs and inject with 2 mL Prostaglandin (PG).
3. **Day 9:** Inject with 1 mL Cidirol.
4. **Day 10:** AI both am and pm.

Synchronisation with blanket insemination

The same program as above, except that all cows are AI'd irrespective of whether they appear to be on heat. This can be costly if you are using expensive semen.

> Research from Virginia Tech (US) shows that the highest pregnancy rates occur when animals are joined five to 12 hours after the first signs of oestrus, which is defined as the 'red zone'. If you breed at 12-hour intervals (e.g. 6 am and 6 pm), you will always hit the red zone.

Semen collection

A lot of stud producers collect semen from their most valuable bulls as an insurance policy. If his genetics are rare or you wish to sell the bull but keep a semen share, an on-farm semen collection is a simple and relatively cheap option. Semen collection is performed by electroejaculation – a mild electrical stimulation, when collected on our farm. Remember that you will need somewhere to store the semen once it is collected.

Semen to be used in export must be collected at a registered and licensed artificial breeding centre. Semen to be used for sale in Australia and/or on farm can be collected by a number of technicians who specialise in this field.

Embryo collection

Embryo transfer is the collection of embryos from the reproductive tract of a donor cow; these are then implanted in numerous recipient cows. Embryos can be implanted fresh, or frozen for future use or export. The donor cow is programmed to super-ovulate.

Many claim that embryo transfer is the quickest and most cost-effective method of increasing the rate of genetic improvement within a herd but be warned – it can be expensive and often disappointing. Some cows can flush eight to 10 embryos and others absolutely none. It is time-consuming as well. Once the embryos are removed or 'flushed' then they have to be implanted into a recipient and there are inherent problems in this. Not all embryos will take (remain inside the recipient). The industry standard is only 50%.

We really only use embryo programs for export purposes now, as it is cheaper to transport embryos than live cattle to foreign countries and it reduces quarantine costs. Also, the embryos are able to adapt to foreign environments first in the uterus and then while suckling a recipient cow indigenous to the area.

Pregnancy testing

Whatever method you use to achieve conception, you may at some stage want to see whether your animal is in calf. This is referred to as PTIC (pregnancy tested in calf). In the US, the terminology is that if an animal is pregnant she is 'bred'.

Pregnancy testing allows you to monitor your reproductive efficiency and detect any problems early in the breeding cycle. It is a very expensive process to run cows that are unable to get into calf . They are sometimes referred to as freeloaders! They cost money in the grass they eat and the vaccines and treatment they receive, and you'd do better to replace them with fertile females.

There are a number of different methods of pregnancy testing.

- **Palpation**: This involves using a gloved hand and arm to palpate the reproductive tract of the cow through the rectal wall. A veterinarian or someone who has attended a course to learn the skills can do this. Some technicians are very experienced and can give you an estimate of how many months in calf the cow is, to within a two-week period. Those who do it the most often are the most accurate. Six weeks is usually the earliest stage at which an experienced person can identify pregnancy. It can be expensive if it's just for a few head and entails a vet but results are immediate. A friend, Sandi Jephcott, has preg tested up to 1500 head in one day!

- **Blood tests**: In Australia we have a wonderful system called BioPRYN, which allows you to take a blood sample from under the tail of the cow. It's simple and inexpensive and you can do it yourself. It is perfect for small acreage holders who may want to test only one or two head. The blood samples are posted in the mail and results are emailed within a week. Positive results can be obtained from 30 days.
- **Ultrasound**: The methods of ultrasound include:
  - introducing the ultrasound transducer into the rectum by hand;
  - having the ultrasound transducer on a pole so no hand is introduced;
  - placing the ultrasound transducer on the flank of the cow.

Advantages of this method are that you can visualise the foetus and sex it within a certain age range. Ultrasound can detect pregnancy with a high accuracy from 26 days. It can be more costly. I am a fan of the ultrasound on the pole and it's a great way to sell stud stock if you guarantee the sex of the calf.

11

Nutrition

This topic encompasses an enormous amount of information so this chapter covers only the elementary requirements for a basic understanding. For the majority of small acreage owners, providing there is plenty of green pasture (and I emphasise the words *plenty* and *green*), your beef cattle should obtain enough protein and energy from the pasture. This can change if there is drought or severe frosts or if you are overstocked, which will require you to either destock or begin supplementary feeding. Depending on soil fertility, species of pasture (introduced or native), environment (tropical, temperate, season) and physiology of the animal (pregnant, growing, lactating), plentiful green pastures may still result in mineral deficiency, toxicity or imbalance.

The aim is to provide adequate feed to keep your cattle healthy, fertile and productive. At different times of the year cattle will have different requirements; for example, during cold and windy weather cattle need a 20% increase of food. Animals that are just maintaining their weight, such as dry (non-lactating) mature cows in early to mid pregnancy, require less energy and protein than late pregnant or lactating cows or young growing calves.

For the inexperienced, there are few basic principles you will need to know.

- **Feed composition**: Cattle need a balanced diet consisting of adequate energy, protein, minerals, vitamins and roughage. Brown dry grass is *not* enough. You will need to know the energy and protein levels of the feeds and pastures that you are using. Your local Department of Primary Industries and the Meat &

Livestock Australia website can be of assistance in this area, as can the nutritionists or your feed supplier (NB: Ensure your feed supplier is reputable and is not just trying to sell a product irrespective of your situation.)

- **Quality**: Always use good-quality supplementary feed that is not dusty, musty or rotten (mostly due to exposure to moisture) as this can limit production, cause serious illness and in some cases death. Appropriate storage of feed will also maintain its quality.
- **Quantity**: Monitor stocking rates through comparing rainfall requirements for your district with current rainfall patterns and learn how to assess pasture requirements until the next rainfall event (pasture budgeting). It is very easy to overstock on a small acreage. Overstocking can be very expensive both in terms of supplementary feeding and long-term degradation of land and pasture species. Contact your local DPI regarding how to monitor stocking rates. It may be a lot cheaper to sell or put cattle on agistment than to feed them through a drought.
- **Observe your animals:** Know how to assess body condition in an animal and what is normal and abnormal behaviour. Use visual appraisal to assess their condition (Table 11.1). Lactating cows are usually the first to be affected by deficiencies in energy as they are using all their energy to produce milk. If they are not receiving enough high-quality feed then a beef animal of average milking ability will start to lose condition, as its daily energy requirements are 30–50% higher than that of a heifer or cow of similar live weight without a calf. Weaned calves are also easily affected: they are restricted as to how much they can eat so the energy levels in their diet need to be more concentrated. Calves also need higher levels of protein, minerals and vitamins for growth.
- **Start slowly**: If you have to supplementary feed then start slowly so the animal's digestive system can adjust from 100% roughage (pasture/hay) diet to a diet containing starch products such as grain (remember: pellets also contain grain) or molasses. Supplementary feeding can be in the form of hay or silage, grain mixtures, pellets, loose mineral supplements, lick blocks or molasses-based products. Hay or silage is mostly used during drought or when cattle are locked up; grain mixtures and pellets are generally used to supplement pastures for cattle that have high energy needs such as lactating cows and growing cattle, or cattle that are being fattened for slaughter or showing; loose mineral supplements, lick blocks and molasses-based products are used to eliminate mineral and vitamin deficiencies and for protein delivery when there is plentiful dry pasture available. Most beef cattle on pasture will need to consume at least 30% roughage when the feed is dry and up to 70% roughage when the feed is fast-growing and green.

Table 11.1: Body condition scoring system for beef cattle

Fat score	Description	Appearance
1 Very lean 0–2 mm	Angular and narrow hips. Pins, backbone and ribs very prominent. No filling in brisket, cut up flank.	
2 lean 3–6 mm	Hips, pins, backbone and ribs obvious but not prominent, flank still cut up but a little filling in brisket.	
3 Medium 7–12 mm	Moderately rounded appearance. Hips, pins, backbone and ribs lightly covered, flank filled to about stifle joint. More filling in brisket.	
4 Fat 13–22 mm	Smooth covering over hips, pins, backbone and ribs to give well-rounded appearance, particularly near tail head. Flank filled to below stifle joint. Brisket full.	
5 Very fat 23–32 mm	Heavy fat covering, starting to be lumpy over hips, pins, backbone and ribs. Pronounced fat deposits near tail head. Distended brisket.	
6 Extreme 33+ mm	Very heavy and lumpy covering over hips, pins, backbone and ribs. Very pronounced ('boxing gloves') near tail head. Very distended brisket.	

Tip: Get your soil tested to determine what deficiencies, if any, are present. These can be rectified with pasture fertilisers or by supplementary feeding of livestock. There is no point putting out salt blocks if you have an excess of salt, as it's a waste of money. The soil analysis will allow you to replace only what is necessary. There are myriad products which can be tailor-made to your specific requirements.

Understanding the terminology

Rumen: Cattle are ruminants. They have four stomachs, one of which is the rumen. The rumen contains billions of microorganisms that convert pasture of various quality into energy and protein that can be utilised by the cow.

Energy (ME): This is expressed as the metabolisable (useable) energy per kilogram of dry feed. Each unit is measured in megajoules (MJ). Energy is available from fats, carbohydrates and true protein. Energy requirements vary greatly depending on the animal's stage of production.

Protein (CP): Expressed as % crude protein. Younger animals need a higher level than older animals. Protein deficiency reduces feed intake and therefore reduces availability of energy to the animal. Protein stimulates activity of the rumen microorganisms that in turn stimulates intake of dry pasture and therefore increases energy intake. Protein is derived from two sources: grain and protein meals, both initially derived from plants and non-protein nitrogen such as urea, which is then converted to microbial protein by the ruminants.

Dry matter (DM%): This is the weight of a feedstuff after all the moisture has been removed by drying in an oven for a set period of time, as a % of the original weight including water. Depending on the quality of the diet, the age of the animal, the breed and the fat cover on the animal, cattle will eat 1–3% of their bodyweight as dry matter intake (DMI) per day. Cattle eat on a dry matter basis and according to energy content of the feed so the lower DM of feed, the higher the intake and the higher energy content of the feed, the lower the intake. Grain is usually 85–90% DM, green pastures are 10–30% DM, dry pastures can be up to 50% DM, hay and straw are 70–90% DM and silage is 20–40% DM (ideally 35% DM).

Digestibility (DMD): Not all parts of feed are digestible. Good-quality feeds have a digestibility of 70–90%. For example, most good-quality pasture is about 80% digestible so for 10 kg of grass eaten, 7 kg is used and 3 kg is expelled as dung. The dung is usually finely textured, sloppy and lies flat on the ground. Poor-quality feeds, such as dry coarse grass or stubble, have a digestibility of 40% and result in coarse-textured dung, which is mound-shaped. Poor-quality feeds take longer for the rumen to process.

Minerals and vitamins: These are essential for healthy animals and optimising growth.

Fibre: Fibre is the fibrous, less digestible part (cellulose and lignin) of the diet. It is more accurately measured as neutral detergent fibre (NDF). In order to maintain a healthy rumen, cattle need a minimum level of NDF in their diet. In particular, they need effective fibre which has a minimum length of 5 cm. Effective fibre is needed to scratch the rumen walls and stimulate rumen movement. On the other hand, when NDF gets higher than 40% in young cattle, it can limit intake.

The food process

Cattle are ruminants. They have a four-stomach digestive system (rumen, reticulum, omasum and abomasum), which allows them to handle fibrous feeds that other animals cannot. Milk is their sole diet for the first month, after that they start to eat grass. This activates their rumen. They eat the pasture or hay by chewing it quickly and mixing it with large amounts of saliva. Some time later this cud is regurgitated and each ball is chewed 40–60 times then reswallowed. Fermentation in the rumen produces energy and protein that can be used by the cattle. From there the fibre is moved through the process of all four stomachs.

Nutritional requirements

The importance of good-quality pasture is that not only is it the cheapest source of feed, it is also essential for growth and production. Assessing whether cattle are

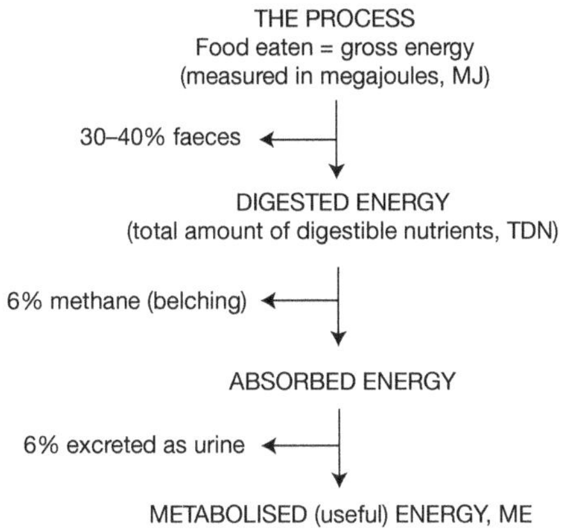

The food process. Source: *Keeping a Cow.* NSW Agriculture.

receiving adequate energy and protein from their diet is a combination of knowing the requirements of the animal (Table 11.2), energy and protein levels in the diet (Table 11.3) and DMD and NDF of the diet which affects DMI. If cattle are not receiving adequate protein and energy and you wish to maintain them in their current body condition, you will have to supplementary feed.

Determining feed requirements

The type of supplement and how much to feed will depend on the nutrients needed and if you want to maintain the animals' weight, minimise their weight loss or produce weight gain. In assessing your options you will also need to consider the

Table 11.2: Energy and protein requirements of beef cattle

Type of livestock	Approximate minimum ME concentration of diet (MJ/kg DM)	Minimum protein level of diet (%)	Daily weight gain (kg per day)
Calves @80–100 kg	9	13	0.5
Calves @80–100 kg	13	18	1.0
Calves/weaners @100–200 kg	9	12	0.5
Calves/weaners @100–200 kg	12	13	1.0
Steers/cows @300 kg	5	8	Maintenance
Steers/cows @300 kg	7.5	10	0.5
Steers/cows @300 kg	10	13	1.0
Steers/cows @400 kg	5	8	Maintenance
Steers/cows @400 kg	7.5	9	0.5
Steers/cows @400 kg	10	13	1.0
Pregnant cow (last trimester)	7	6	
Lactating cow and calf	10	10	

Source: DPI Victoria, Information Note AG0374.

Table 11.3: Average nutrient analysis of some commonly used feeds and hay

Feed	Approximate metabolisable energy (MJ/kg DM)	Crude protein (%)
Soybean meal	11	45–48
Cottonseed meal	10	34–36
Whole cottonseed	14	25
Copra meal	12	20
Lucerne hay	8.5	18
Barley grain	12	12
Rhodes hay	8	12
Sorghum grain	12	12
Oats grain	11	11
Kikuyu hay	8	9
Sorghum hay	8	8
Oaten hay	9	7
Wheaten hay	8	6

This analysis will vary within feeds depending on seasonal condition during growth, varieties and fertiliser application.

cost, both financially and in human resources needed, as delivering supplementary feeds is often labour-intensive and costly.

A recent research project found that a high-protein diet in the first 90 days of pregnancy can have a significant impact on embryo survival, foetal growth and subsequent development of the offspring. The aim of the project was to determine the economic value of supplementary feeding and when it was most beneficial.[4]

Before you begin supplementary feeding, to avoid wastage and excessive costs, you will also need to know:

- the live weight of your animals;
- the body score condition of your animals;
- the quantity of feed in the paddock;
- the sex and age of your animals (steer, heifer, cow, bull);
- the pregnancy and lactation status of the animals;
- breed (frame score or mature weight of the animals, e.g. Charolais, Lowline, Brahman).

This will enable you to sort the cattle into different feed and management priority groups, e.g. cows with calves, first calf heifers, weaners, steers, dry cows and bulls.

As explained in the introduction to this chapter, there are a number of ways to supplement deficiencies in your animals' diet:

- lick blocks;
- hay;

Table 11.4: Maintenance feed requirements with no additional pastures

Type of stock	Total daily energy required (MJ ME)	Kilograms grain only (12 MJ ME)	Kilograms hay only (8.5 MJ ME)	50:50 grain:hay mix	80:20 grain:hay mix
Calves/weaners @100–200 kg	25	2.5	3.5	3.0	2.5
Yearling @200–300 kg	35	3.0	4.0	3.5	3.0
Adult dry animals @300–400 kg	45	4.0	6.0	5.0	4.5
Dry cow in late pregnancy	50–65	5.0	8.5	7.0	6.5
Cow (over 400 kg) and calf	85–95	Nil	10.5	9.0	6.5 kg/ 1.5 kg hay*

* Too much grain affects milk levels. During cold times these levels should be increased by an additional 20% using hay.
Source: Adapted from McKiernan B, Prime Fact 276, DPI NSW.

- molasses or molasses-based products such as Anipro and Molafos®. Most molasses-based products contain urea in various levels. Some products, such as Anipro, control intake by the use of acids;
- grain;
- protein meals.

Table 11.4 gives an indication of how many kilograms of supplementary grain and hay is required for animals at different stages of development and production. These are for maintenance requirements only; if you want the animals to put on weight then the rates would have to be increased.

Tables 11.2–11.4 show that as some hays have a protein level of only about 8% this would not provide enough to feed calves, weaners and lactating cows. Protein levels would have to be supplemented.

Tip: Early intervention is by the far the cheapest option. It is much cheaper to maintain an animal's condition than it is to have to feed to put condition back on.

Q. What sort of hay do you feed?

A. We normally feed forage sorghum or good-quality 'grassy' hay. The only time we feed prime lucerne is a biscuit each on the morning of show day to give the cattle a bit of extra fill. If used as the only source of supplement, prime-quality lucerne hay will produce an imbalance in protein and energy, which can cause diarrhoea.

Q. How much hay should I feed?

A. Medium- (6%) to high-protein hays (16%+) can be fed at a rate of 1–2 kg per head per day when there is plenty of paddock feed available. A small square bale weighs approximately 25 kg.

Tip: Remember to pick up all baling twine as ingestion can cause choking and death.

Q. What supplements, if any, do you feed?

A. Laidley is situated in the subtropics, which has a summer-dominant rainfall. Tropical pastures are dominant in these areas, which is excellent for the high-rainfall summers; however, these grasses do not tolerate frost so lose quality and quantity during the dry frosty winters. Hence, during the winter we feed hay in the form of home-grown baled forage sorghum and Anipro. The hay provides the roughage and some energy, the Anipro (a molasses-based supplement) provides all the necessary vitamins and minerals as well as non-protein nitrogen (urea).

Anipro and other similar supplements are *not* sources of energy – the molasses is diluted and only used as a suspension tool for the urea, mineral and vitamins. I like the Anipro form of supplement for ease of use: the supplier provides the troughs, comes regularly to refill them and the mix can be adjusted to suit the type of stock you are feeding. You are charged per litre and Anipro is considerably more expensive than buying straight molasses. There are a number of similar molasses-based products available. Some are cheaper than others and some can only be purchased in bulk, which often does not suit small acreage farmers.

Anipro is a safe method of feeding urea as it suspends the urea evenly throughout the molasses and uses hydrochloric acid to control intakes. Urea is a cheap and effective source of non-protein nitrogen which increases intakes of low-quality dry pasture. Unfortunately, urea is also toxic and will readily kill animals if the rumen is not adapted and cattle eat too much of it.

Simple rules for feeding urea is that it should be fed only when there is ample dry feed available, for example during winter in the tropics or subtropics. Never feed a urea supplement in a severe drought situation – cattle need a roughage and energy source *not* a urea supplement in a drought. Introduce urea before the feed deteriorates severely and never let the cattle run out of urea during the period it is being fed. If cattle run out of urea, they will overeat when it is reintroduced. If urea is fed as a dry supplement, blocks or home-made molasses mix, ensure the supplement is mixed well. If

The Anipro troughs restrict access to calves once the level gets below halfway and are only suitable for adult animals, so you need to make sure that calves have access as well.

it rains, tip the supplement out or remove the water as the water will contain highly concentrated dissolved urea which will readily kill cattle.

Q. Does Anipro provide enough nutrition for dry cows grazing irrigated short green-pick during a drought?

A. (This response was supplied by Anipro's technical manager.)

Let's run a nutritional forensic analysis. First, what are the cow's requirements?

Most important is her feed intake. She must eat to fill her rumen every day. Her potential intake increases with nutrient demand. A lactating cow can eat a lot more than a dry cow. Anipro offers no dietary bulk, so it will not contribute to daily intake and can be counted as zero in the diet. However, Anipro, like most protein supplements, will stimulate the appetite. Just like us taking a few vitamin tablets does not reduce hunger, the nutrients actually aid digestion, we may be able to eat more.

The second priority is energy. Energy for cattle is measured in megajoules (MJ). Plant material is measured in dry matter. For maintaining bodyweight, a dry cow needs 40–60 MJ each day (depending on her frame size and live weight). Young grass has about 10–12 MJ/kg of dry matter. However, it is about 90% water. So a cow must eat 40–50 kg of fresh green grass each day to get enough dry feed to fill her rumen and get enough energy. That's a lot of grass to grow. Older grass has slightly less energy (8–10 MJ) but is a lot higher in dry matter (40–60%). So a cow would need to eat only 12–15 kg of this pasture. Anipro has only 4 MJ and intake is only 0.4–0.5 kg/day. This is not enough energy to worry about.

The third priority, but the easiest to correct, is dietary protein.

Protein nutrition for cattle is a complex topic. To keep it simple, let's assume that a dry cow needs about 600 g of protein in the diet each day. This protein must be in a form that the rumen bugs can use as fuel. So the day's pasture intake should have 7–10% useable protein. When the useable protein concentration is low, the animal's system suffers. Rumen bugs stop reproducing, which slows digestion. Slow digestion means reduced appetite and therefore less nutrient intake.

Anipro can now help. At best, Anipro offers 160 g of protein in a highly useable form. This protein, along with the trace minerals, vitamins and the added rumen modifier Flavo, will stimulate rumen bugs to reproduce, work harder and get the system firing. Appetite is stimulated and as a consequence nutrient intake should also rise. Problems with this system may be noticed in two cases: when the availability of dry matter is limiting, and when the digestibility of the dry matter is low. Remember, the first priority for cattle in a drought is maintaining feed intake. If the offered hay is low-quality, then the fibre cannot be broken down by the bugs and the rumen stays full. Intake is again depressed and cattle can look to the Anipro trough for sustenance.

Anipro excels as a production stimulant as it can take cattle to a high level of nutrition. It does work in a drought, but only if there is enough digestible dry matter for the cattle to eat every day.

Q. What feeds do you feed your show team? How long do you feed them for?

A. Feeding cattle for show preparation is much more intensive and requires more skill than feeding for maintenance. Animals need to be presented in show condition without being overfat. We feed each animal individually on a daily basis, split over two feeds, and at the same time each day. Cattle are weighed and recorded every week to assess their weight gain. Young bulls require much more feeding than heifers as it is virtually impossible to get young growing bulls overfat but quite easy to do with small breed heifers.

We feed our show team for a period of eight to 12 weeks leading up to a royal show. We start with a very small amount of stud mix (0.25% of body-

Table 11.5: Available metabolised energy of raw and micronised grain

Grain	Raw (MJ/kg)	Micronised (MJ/kg)
Maize	14.3	15.7
Wheat	12.9	14.4
Barley	11.5	13.7
Sorghum	11.36	13.63

Source: Mi-feeds.

weight) and slowly increase the amount (1–1.5% of bodyweight) along with the various supplements. The cattle are also fed hay (forage sorghum or grassy hay) for roughage. It will take three weeks before additional feeding will start to take effect or increase weight significantly as the microbes in the rumen need time to adapt.

We have used a pre-mixed stud ration from Bremer Stockfeeds for the past 10 years with great success. If micronised feeds were available then we would use them. The process of micronisation basically cooks the starch in the grains, thus making it more digestible for the animal. The grain is first steamed, then heated to a high temperature and then rolled. It produces significant increases in feed efficiency (see Table 11.5). The utilisation of grains being fed can be assessed by the amount of grain seen in the faeces. A lot of grain in the faeces indicates the grain is not being processed adequately to allow digestion by the microbes and enzymes in the digestive tract. Some grains, such as sorghum and wheat, are more difficult to process for efficient utilisation.

We add a number of supplements to our feeds. They include:

- black sunflower seeds for the oil content to give shine to their coat;
- micronised soyabeans to add fat and bypass protein (high-quality protein that is not affected by the microbes in the rumen and is absorbed directly in the small intestine);
- copra to add fat (energy), protein and fibre and give shine to their coat.

Q. I have been told to be careful to avoid grain poisoning. What is grain poisoning?

A. When grain is introduced to the cattle's diet it needs to be done slowly otherwise their digestive system will be upset. If cattle that have been grazing pasture suddenly eat high levels of grain without additives such as sodium bicarbonate, bentonite and ionophores/rumen modifier (e.g. rumensin) to stabilise the rumen, fermentation of the carbohydrates occurs, lactic acid is produced thus reducing the pH in the rumen to <5, and acidosis or grain poisoning, dehydration and death may occur. If you are going to give your cattle grain, restrict access to it and make sure they have access to plenty of roughage such as hay that will help the digestive process. The risk of acidosis is significantly reduced if the hay is mixed with the grain rather than the grain being fed in a self-feeder and the hay being fed in a hay rack. Alternatively, cattle should be let in on the hay rack and allowed to partially fill their rumen with hay before they are given access to the grain supplement. Most grain poisoning occurs when cattle have access to self-feeders, they are overfed grain in troughs, adequate reasonable quality roughage is not provided, grain is fed to hungry cattle or they break into the feed shed.

We lost some stud cattle one year when I was away. Dairy meal was ordered and delivered into the self-feeder instead of the usual stud mix. The dairy meal had higher energy levels than the stud mix, no sodium bentonite was included and the self-feeder was not locked down as hard as it should have been (some of the cattle had previously been supplemented so were very aggressive at the feeder). We suddenly had about six cases of severe grain poisoning. We lost three head and managed to save three by using 500 g of bicarb soda dissolved in 2 L of water (1 g bicarb soda per kg live weight) used as a daily drench, penicillin injections and vitamin B1 injections. Cattle with severe acidosis should have restricted access to water, as they will bloat on water, causing respiratory arrest. Severe cases need veterinary assistance.

Tip: Always keep feed shed doors shut and locked to prevent any accidents.

Q. Can you tell me the best method to feed supplementary grain?

A. For the inexperienced, I would first recommend:

- a pre-mix, such as pellets or ration, from a reputable supplier rather than mixing your own. Make sure it has a rumen modifier, bentonite or sodium bicarbonate added. They help prevent grain poisoning;
- if you are mixing your own grain then you will need to add either 3% bentonite or 1% bicarb soda to the mixture to reduce problems of grain engorgement. These additives will reduce but not eliminate problems. Appropriate feeding management will eliminate grain engorgement.
- start slowly with a little and then over time gradually increase the amount;
- ensure prior to and during grain feeding, cattle have access to hay or pasture. Do not feed grain mixtures to hungry cattle;

- put the feed either in a self-feeder, which can be adjusted to regulate the flow of grain, or in a trough. Make sure there is enough space for all animals, as the bossy ones will dominate. Never put grain straight onto the ground, as there will be too much wastage;
- if you are using a grain bin make sure the gap is as small as possible for the first week or so then slowly lift it, allowing the cattle to have access to larger amounts.

Q. Do you use any trace elements? I have been told there are some good injectables available.

A. Yes, we use Multimin when weaning and in times of drought. It is a chelated trace mineral injectable source of zinc, manganese, copper and selenium. Multimin is recommended as ideal for topping up trace minerals prior to critical events (e.g. joining, calving, marking, weaning) when demand may increase.

Q. I noticed the female not eating much. Sat and watched her for half an hour; she didn't eat. She was just looking around. Is this normal? Excuse my inexperience.

A. Contented cows don't eat all the time. They normally rest during the hot part of the day to conserve energy and to chew their cud. Chewing their cud is important as, when they do this, they secrete saliva. This saliva contains a natural antacid which helps to buffer the rumen or first compartment of the stomach. Proper buffering of the rumen allows a cow to digest forages better and to eat more feed, which helps her produce more milk. Cows spend nearly eight hours out of every day chewing their cud.

12

Healthy soils, healthy pastures, healthy cattle

My experience is that most people on smaller acreage embrace the idea of sustainability. Not only do producers wish to preserve the land for future generations, but consumers are becoming more aware of our food's integrity and more concerned about its quality. This chapter explains the notion of achieving balance in our land management techniques and practices that include the soil, plants and livestock. Only the basic concepts are covered here as there is a wealth of information available and many books and articles relating to this subject. Healthy soil results in better pasture quality, which in turn produces healthy cattle, allowing you to produce more beef per hectare. By taking a holistic approach, rather than a band-aid approach, you will achieve a far better outcome in a much healthier ecosystem. There are a number of methods to improve your soil and your pasture. You may choose to use fertilisers, you may choose to be totally organic or you may choose to go that one step further and be biodynamic. Gaining certification for the latter two requires a longer time-frame, a lot more effort and a total restructure of your farming practices but the rewards are certainly worth it.

Healthy soils

The concept of soil health has three foundations – chemical, physical and biological. Maintaining the physical and chemical (including the pH and nutrient levels) assists the biological. And in return the biological maintains the integrity of the physical and chemical aspects. Before you start on any program to improve your soils and your pastures, it is advisable to get a soil test done. Soil testing identifies potential toxicities and nutrient deficiencies. Overproduction of our land has resulted in erosion, compaction, acidification, rising salinity, pasture degradation and, most importantly, loss of organic matter and vital nutrients. Producers have to use more chemicals to control weeds and more fertilisers to make the pastures grow. Both these practices are expensive and unsustainable. A renewed focus on soil health will not only improve the nutritional quality of the food you produce, it will also:

- increase agricultural productivity without high input costs;
- extend the growing period and boost water-holding capacity;
- improve downstream water quality;
- boost soil fertility;
- enhance ecosystem services such as clean water and air. A healthy soil is teeming with a host of organisms which improve the soil structure, which in turn improves the soil's ability to drain and retain moisture.

There are a number of practices that producers can engage in to improve the health of their soil:

- don't overstock;
- add organic matter such as well-rotted manure;
- grow green manure crops;
- practise cell grazing;
- avoid overworking the soil and use zero till practices;
- reduce soil compaction;
- apply organic fertilisers;
- use dung beetle friendly products.

The P-Tool: a decision support tool

The P-Tool (available on the MLA website) outlines five easy steps that allow producers to understand the value of soil testing and how to use soil test information to plan fertiliser and livestock investments. The information package provides a better framework for understanding and planning the use of phosphorus (P) fertilisers. The tool is intended to assist producers to determine suitable levels of P fertilisation for temperate pastures grazed by beef cattle on acid soils in southern Australia.

Healthy pastures

You need good-quality pasture for your livestock to grow. In fact, I always say I grow grass first and foremost and beef second. I can't produce beef if I don't have

Ag Solutions AUSTRALIA
Helping Australia Grow

Soil Test Results

Date: 03/10/2011	Customer ID: 11791
Margo Hayes	Block Identification: PADDOCK 3
	Phone Number: 5466 7266
221 Mulgowie Rd, Thornton via	Fax Number: 5466 7288
Laidley QLD 4341	Crop Type: pasture

Element or Category	Your Level	Optimal Level	Nutrient Status (Low – Marginal – Optimal – High – Excessive)
pH *1:5 water*	8.4	5.7 – 7.3	
Elec. Cond. *1:5 water*	0.13 mS/cm	0.06 – 0.5	
Nitrate (N) *water-soluble*	34 ppm	20 – 55	
Phosphate (P) *BSES*	773 ppm		
Phosphate (P) *Colwell*	419 ppm	35 – 90	
Potassium (K) *Am Ac*	1008 ppm		
meq	2.58	0.35 – 0.8	
cations	6.4%	2.8 – 6.0	
Calcium (Ca) *Am Ac*	4805 ppm		
meq	24.03	8 – 16	
cations	59.4%	63 – 78	
Magnesium (Mg) *Am Ac*	1650 ppm		
meq	13.64	2.0 – 5.3	
cations	33.7%	14 – 22	
Sodium (Na) *Am Ac*	46 ppm		
meq	0.20	0.25 – 0.55	
cations	0.5%	1.3 – 3.5	
Sulphate (S) *KCl*	32 ppm	30 – 50	
Zinc (Zn) *DTPA*	8.9 ppm	1 – 10	
Copper (Cu) *DTPA*	2.48 ppm	1 – 10	
Manganese (Mn) *DTPA*	9.83 ppm	5 – 55	
Iron (Fe) *DTPA*	396 ppm	5 – 170	
Boron (B) *$CaCl_2$*	0 ppm	0.1 – 1.1	
Organic Carbon *W&B*	2.35	2.4 – 6.0	
Chloride (Cl) *Water-soluble*	60 ppm	<150	
Ca:Mg Ratio	1.76	2.7 – 5.0	
Cation Exch. Capacity	40.44	Clay	

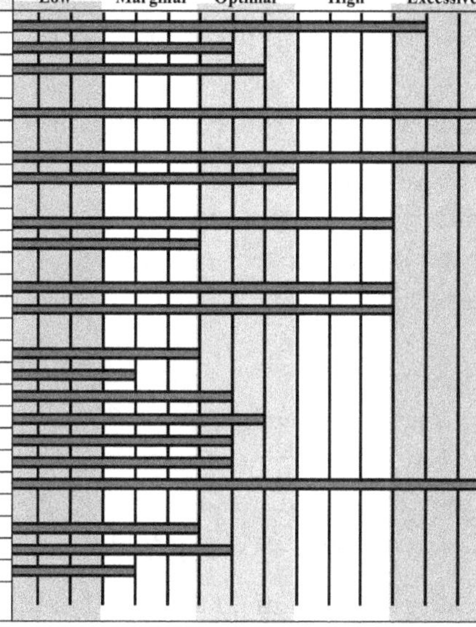

Recommendations:

- Top dress with **NatraMin Cal-S** @ 300 - 400 kg/ha, there is no withholding period for livestock. Due to the elevated levels of Magnesium and impacts on soil structure, repeat this application annually as part of a soil improvement program.
- There have been some changes to this paddock since 2009 soil test. There have been improvements with increased available Phosphorus, Nitrogen and a small lift in the trace mineral levels as well as a slight improvement in the Ca:Mg Ratio.
- The effects of a soil becoming waterlogged or anaerobic are many. The severity of these effects depend on the duration and frequency of the waterlogging event, with some of the effects being; loss of top soil, dramatic changes to soil biology, changes to soil structure, increases in compaction, loss of soil nutrients, changes to the availability of soil nutrients, loss of soil organic matter, yield loss and increased soil salinity.
- NatraMin Cal-S acts as a fertiliser and soil conditioner with a bio-stimulant that acts to increase microbial activity in the soil. It also provides higher levels of Calcium and Sulphur to assist in the balance of cation exchange, which will improve the Calcium: Magnesium ratio which is also affecting soil structure.
- Soil aeration and rotational grazing, where appropriate, also compliments NatraMin to improve soil structure and pasture growth.
- NatraMin stimulates worm and microbe activity in your soil. Soil microbes produce their own weight in humus every day (assisting to floculate soil structure and build soil carbon) and they also increase nutrient availability. Organic carbon levels are low (3% is desirable); hence any opportunity to return organic matter to the soil (like rotational grazing) and enhance biological activity should prove beneficial over time.
- By re-mineralising the soil we are assisting in getting the balance right between structure, biology and nutrition. NatraMin blends assist in releasing locked up minerals and nutrients to make them into more available forms for plants. NatraMin Cal-S also contains a broad spectrum of trace elements to cater for most pastures.
- NatraMin blends also contain a high level of Silica. Silica plays a role in regulating plant uptake of Aluminium, Manganese and Iron (no need for liming) It has a major influence on the absorption of macro and micro nutrients. It strengthens the cell walls of plants reducing damage from frost. There can be up to a 50% increase in chlorophyll levels in plants and up to 100% increase in root thickness.
- Due to a world shortage of the chemical required to do Boron extraction tests for soil, sap and water, we are not able to test for Boron. We apologise for any inconvenience that this may have caused, however this is totally out of our control.

A soil test showing recommendations for improvement.

the pastures to feed my cows. Most people need to know the size of their property and their paddocks. We have drawn up a paddock plan that details the shape and size of each paddock. This helps to calculate stocking rates and enables us to budget for available feed. There is a very useful app called Measure Your Land that can help you do the same.

Like your soil, your pastures are an investment that needs to be protected. The roots of your pasture are one of the best tools available. They are a primary source of soil carbon and improve the texture of your soil, enabling better drainage, so this is the best place to start.

If your pasture is in poor condition then you may have:

- little or low ground cover;
- reduced rain absorption;
- greater soil erosion;
- poor soil structure;
- poor root growth;
- reduced fertility;
- minimal organic matter;
- inability to absorb nutrients;
- high percentage of weeds;
- increased incidence of soil acidity.

If your pasture is in good condition then you will:

- have 100% ground cover;
- maximum rain absorption;
- minimal erosion;
- better soil structure;
- better soil density;
- increased organic matter;
- increased nutrient cycling;
- plant biodiversity.

Good grazing management strategies directly impact upon pasture productivity.

- **Recovery time**: Pastures must be allowed to recover after grazing to restore plant energy reserves and promote growth of roots to a deeper level. Deep root structure is an important source of organic material for soil organisms.
- **Overgrazing**: Overgrazing pastures is detrimental in many ways. It can cause compaction, which prevents the pastures being able to develop adequate root growth. It causes degradation of the available pasture, allowing weeds to flourish and sometimes take over. It can expose the soil, so a sudden deluge of rain can cause massive erosion of topsoil and pasture.

- **Avoid burning**: Many farmers in my valley burn their paddock every spring to try to control weeds and lantana. Burning just destroys the available organic matter.
- **Nil tillage**: Mechanical soil tillage should be avoided as it affects the soil's structure, disrupts the habitat of soil organisms (which reduces their presence) and reduces the organic matter.
- **Fertilisers**: A soil test will identify your soil deficiencies, which will allow you to supply all the nutrients and trace elements required to maximise the benefit from fertiliser applications. For example, phosphorus applications in sulphur-deficient soils will not have any effect unless sulphur is also supplied. Poor timing can also greatly reduce the response; for example, applying phosphorus to wet and cold soils is less effective. Supply of essential trace elements such as molybdenum may be required. If deficient, the nitrogen-fixing capability of plants is reduced and your pastures will not grow as well.
- **Lime**: Applying lime is the only recognised method of correcting soil acidification.[5]

Understanding the principles of rotational grazing

The quantity, quality and utilisation of pasture and feed grown can greatly affect your productivity and profitability. Rotational grazing is based on the principle of moving stock around a number of paddocks, grazing each for a fixed period of time, to improve all three factors.

There are a number of advantages to such a system. They include:

- greater use of pasture production and increased stocking rates of up to 20%. The reason is that rotational grazing is highly intensive (larger groups on smaller areas) and so reduces patch grazing, i.e. all of the available pasture is eaten;
- reduction of broadleaf weed growth due to the intensity as weeds are eaten or trampled;
- reduction of worm infestations as their life-cycle is broken by cows continually moving onto fresh pasture;
- improved husbandry as stock are viewed more regularly as they are moved from paddock to paddock;
- by grazing small areas intensively then resting them you not only prevent the pastures from deteriorating but also lessen the pasture that is lost to trampling. Cows' feet consume five times more grass than their mouths do;
- longer periods of rest can assist with reducing tick infestation.

A simple way to start is with a four-paddock system. This can be very beneficial for those who want to run all their animals together. The amount of plant regrowth is entirely dependent on the number of stock grazing each paddock and the rest

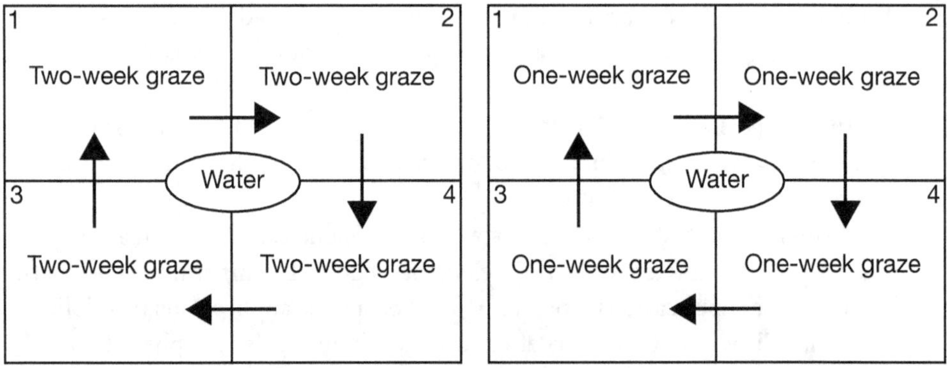

Left: Moderate pasture growth rotation, allowing each paddock a six-week rest period. Right: Fast pasture growth rotation, allowing each paddock a three-week rest period. Source: *MLA Tips and Tools: Getting Started in Simple Time Based Rotational Grazing* (2003).

period between grazing. This in turn determines how much pasture energy is converted into saleable meat. You will need to ensure that each paddock has a water supply for cattle .

In times of moderate pasture growth then the grazing period is greater, allowing sufficient growth in all paddocks. In times of fast pasture growth then the rotation is greater, so that the pasture is not able to get too long and decline in nutritional value – the most productive pasture is that in the short green growing phase with a high digestibility and high-energy value.

Noxious weeds

When you buy your land, use agistment paddocks or move stock into new paddocks for the first time check beforehand for any noxious weeds. Ask your local vet or DPI what poisonous weeds exist in the area. Under most normal conditions cattle will ignore the majority of plants, but being introduced to new paddocks or a lack of alternative feed may prompt stock to eat them.

Many pastures grazed by cattle often contain some poisonous plants. In small amounts, cattle can tolerate many of these plants. Normally cattle avoid many of the toxic plants but if they are grazed extensively or are the only feed available then they can have a disastrous effect. In times of drought, when there is nothing else cattle will eat plants they normally avoid.

Some plants are toxic even in very small amounts. I once leased a paddock that had grass so high I could hardly find the cattle, and within two days I had lost three of my best stud females. The problem occurred very quickly; they were dead within six hours of ingesting this plant. My vet and I walked the paddock, finding a few bushes of green cestrum. A weed I had never heard of or knew what it looked like, and neither did my cattle. Some poisonings, as in the example given, have highly visible outcomes, while others can develop more slowly and are cumulative, with damage to internal organs developing over time.

Meat & Livestock Australia lists the following as examples of potentially toxic plants grazed by livestock in Australia:[6]

- Paterson's curse;
- perennial ryegrass;
- pimelea;
- St John's wort.

Others I would include, as I know of breeders who have been affected, are:

- oleander bush: eating only 10–30 g can be fatal. The dry leaves retain their toxicity. Poisoning often occurs when garden rubbish is disposed of where cattle can get access to it;
- bracken and mulga fern;
- lantana: there are 29 varieties in eastern Australia. The red-flowering and 'Helidon white' bush is toxic, orange-flowering less so and the pink-flowering is toxic only north of Rockhampton. Clinical signs include loss of hair due to photosensitisation;
- mother-of-millions;
- azaleas;
- arum lilies;
- Noogoora burr: same toxin as in green cestrum, acutely toxic to the liver;
- castor oil plant: poisoned animals develop severe diarrhoea.

13

Showing

Showing cattle can be a fun and enjoyable hobby or a way to market your stud, meet other breeders and involve all the family. It can be terribly exciting, especially when you win your first ribbon, but it can also be time-consuming and expensive. To be competitive, especially at a royal show level, cattle have to be in show condition – and that means feeding them correctly and presenting them properly.

I have shown over 50 Grand Champion animals and competed at all the royal shows in the eastern states. In that time I have beaten some pretty serious competition. There is nothing more rewarding than winning the coveted red, white and blue Grand Champion ribbon or an interbreed sash, especially when it is an animal that you have bred yourself. It's also a way of benchmarking against other breeders and against other breeds.

A number of people offer a fitting service: someone who will feed, prepare and show your cattle for you. This can be costly but an experienced fitter can make the difference between first and second place, Grand or Reserve Grand Champion. They know how to present the animal, keep an eye on the judge, parade and stand the animal correctly. A number of breed societies run handling schools throughout the year and these are an excellent way to learn how to prepare animals for the show ring.

I have always employed the services of someone to parade and present my cattle at royal shows. Julie and her husband Travis and I make an excellent team and we work well together. I breed, choose, halter-train and feed the show team while Julie and Travis clip, present and lead them. Their expertise and skills have on many occasions helped my cattle win that coveted Grand Championship ribbon.

If you want to show yourself, then start at local show where it is always much more informal and relaxed. The royal show ring is not the place to begin. Many breed societies provide advice and information on showing, so make contact with them.

Preparation

Planning

Plan your show season in advance. Work out which animals you wish to show, which shows you wish to attend and what classes they will enter. If you're serious about showing, this plan can also affect your joining and calving times for specific animals. For example, if I have a very special heifer that has done exceptionally well in the show ring then I plan her joining so that she will have a calf at foot the following year specifically for the royal shows I wish to take her to.

Halter training

You will need to teach your animals to be halter-trained. This is called being led and tractable. There are a number of videos that do an excellent job of teaching these skills. Most importantly, do it slowly. Don't try to halter-train an animal in one week. Everyone has a different method. My method is based on a reward system, kindness and time.

1 Start them as young as possible. It is so much easier to break in young cattle than adult ones. The younger the better as they will have less chance of getting away from you, as they are less strong and have had less chance of learning bad habits.
2 Take your time. I do it slowly over a four-week period. I begin by feeding them individually for two weeks so they get used to me, being fed and the associated noise of buckets and people.
3 Then I put them in the crush and fit a halter and lead rope. Make sure it is a chained hackamore and not a rope halter. This is most important. Once the animal is tied up the chain will release when the animal comes up to the post or stops pulling away, allowing it to learn right from wrong. Rope halters often don't give, so the animal can't possibly learn right from wrong. Tie them up in the crush or race for 15 minutes with their head held high. This prevents them from being able to thrash about too much and allows them only to try to go backward and forwards, not sideways. Watch them carefully: some animals will try to throw themselves on the ground and you must be able to release them quickly if need be. Learn to do quick-release knots. Talk to them calmly. Tell the cattle they are being good and do everything gently. Animals respond to voice and body movements.
4 After 15 minutes, let them go and leave the halter on them if you are keeping them close by. Give them some hay or grain as a reward. It is dangerous to put

Left: Correct halter for breaking in as the chain will give as the animal comes up to the rail. Right: Incorrect halter for breaking in as the rope will not give or release the pressure.

cattle in a big paddock with a halter on, as it can catch in fences. We leave them in small holding paddocks beside the yards.

5 On the second day, tie them up again either in the race or onto a fence post in a small yard. You may have to put them in the race to attach the lead rope. Tie them up at head height, nice and short. Brush them with a broom. I suggest a broom to avoid any danger of being kicked. Brush on both sides down the legs, up their sides and up their necks. Talk to them as well.

6 Repeat this for the third and fourth day for 15 minutes but no longer. Don't overdo it. They will be sore on their chin from the halter chain by this time. Reward them with feed once you have untied them.

7 Give them a break for the fifth and sixth days. You can leave the halter on but don't tie them up.

8 Seventh and eighth days. Tie them up for 15 minutes. Then untie and gently pull them to one side to face you. Move in the opposite direction, pulling their heads around so they face you. Try to pull them up to you. When they come, give on the rope so the pressure is released and they learn that if they do what is being asked then it won't hurt or be uncomfortable. Keep doing this from side to side for 10 minutes. Finish on a good note and let them go. Reward them with a feed.

9 Ninth day. Repeat the same process but in a slighter bigger yard. It is a good idea to have someone close by to help at this stage. Try walking with each animal and leading it around the yard. If it won't come when you ask then get someone to tap it gently on the backside, while you are pulling.

10 Repeat this process each day, moving into a bigger area until you are confident that you can lead them in a show ring.

Tip: Learn how to tie cattle up correctly with a quick-release knot so if the animal gets into trouble you can undo it quickly.

Education

Before you take cattle to their first show, try to get them used to as many things as possible that they may encounter. Quiet calm animals will not only gain weight faster but will parade better in the show ring. Unruly animals can be disqualified from the show ring.

They will probably be nervous on their first outing so here are a few things to try at home.

- Put a ribbon on their halter and leave it on for a few minutes. That way they won't shy so readily if they are lucky enough to be a place-getter.
- Tie them up and turn the radio up so they get used to noises. Leave the radio on while they are being fed.
- Wash and brush them regularly.
- Put a nose clip on and leave it in for 10 minutes or so.
- Practise using the show cane to make them stand up, so they don't take fright at the stick.
- Wash and clip them so they get used to the idea.

Clipping

You will need someone to teach you how to clip, or find someone who will come and do it for you. Different breeds have different preferred styles so find out what other breeders are doing. Normally an animal will need clipping every four weeks and a slight trim just prior to the show ring.

Hoof trimming

This is a very sensitive subject and, while most people pretend that cattle don't need to have their hooves trimmed, the truth is that you would be hard pushed to find any animal at a royal show that has not been done. This is mainly due to the fact that they have been yard-fed for some three months: because they have been on soft feed they have not had access to paddocks with rocks and gullies that help to trim their hooves naturally. Not all animals will need to be trimmed, especially young ones, but older animals may. It is just like having your fingernails trimmed and is in no way harmful to the animal. The issue is controversial when people use hoof trimming to hide or try to correct a number of serious structural faults.

Entries

Once you have decided which shows you wish to attend, ring the respective show societies and ask to be put on their mailing list for a show schedule. Royal show entries often close two to three months before show dates whereas some local shows take entries up until show day. You will need to make sure all animals are

Cattle are walked into a portable table. This is then turned very gently on its side, causing no harm to the animals, and an angle grinder is used to trim the hooves.

registered, to facilitate entries, as many shows will not accept pending registrations. Many shows also require that you owned that particular animal (i.e. registration was transferred) at least three months prior to show date.

Classes

Find out what the classes are and what animals you will take. Stud stock classes are based on age and normally, providing there are adequate numbers being shown, are divided into junior and senior sections. There will be a junior and reserve junior for both sexes and then a senior and reserve senior champion for both sexes. Out of these two championships a Grand Champion is chosen for both sexes. Some breeds also have a supreme exhibit, chosen from the Grand Champion male and the Grand Champion female. Eligibility dates for each class will be printed inside the entry schedule. Read the schedule rules to find out eligible age of calves, if there is an age limit for senior classes and if you are restricted on the number of entries per class.

> **Tip:** Consider carefully when entering a dry cow in a senior cow class at a royal show. The point of this class is to assess the cow's ability to reproduce and, unless your female is of exceptional quality or very close to calving, the majority of judges will invariably place a cow and calf unit above a dry cow.

Identification

It is a prerequisite for royal shows that the majority of entries must be tattooed in their ears and carry an NLIS device. The relevant breed societies will advise you of their tattoo regulations; for example, some breeds, such as Brahmans, are not required to tattoo.

Nose rings

These are compulsory for male animals over 12 months including some led steers.

Nose clips

These are compulsory for female animals over 12 months.

Show day

Be organised on the day. Most classes start early morning so, depending on the number of head shown, this will mean an early start.

- We wash our cattle the day before so on show day they just need a sponge and blow-dry.

Four months' preparation for a show. Top left: Before feeding began. Top right: Just washed and before being clipped. Bottom left: After being clipped. Bottom right: Show day presentation.

- Pre-fit their leather show halters and name them so they are easily sorted on the day.
- Find out which ring you are being shown in so you can find a tie-up space close by early, before everybody else does.
- Know which animals are going into the ring first, and what the order will be.
- Make sure you are there when the class is called. Shows have a strict timetable and royal shows especially often won't wait for exhibitors that are late to their class. If you have a big team and need some time between classes then let the stewards and judge know, so they can accommodate you.
- Keep an eye on what is happening in the ring. If you have an animal that came second to one that was awarded a championship prize then be ready, as you will be called into the ring and have an opportunity to parade for reserve championship.

What to take

Write a list of all the things you will need. The following may be of some assistance:

- white show coats or breed uniform – until recently the only acceptable dress code in the show ring was a white coat. The purpose was to make everyone uniform with no display of stud logos, and allow the judge to concentrate on judging the cattle not the parader. Nowadays most breeds have their own uniform due to the fact that white coats are very impractical when handling cattle;

- bedding – check to see if the show society is providing bedding for your cattle or if you have to take your own. Royal shows will provide it and include the cost on your entry form, while most local shows require you to bring your own;
- wheelbarrow, rake, broom and shovel for keeping the beds clean;
- rubber mats for the animals to stand on and to prevent bulls digging holes in the bedding;
- show sign;
- brochure stand and promotional material;
- water buckets and feed buckets;
- feed (hay and grain);
- hay nets;
- baby powder – needed to absorb the gel that is used to scan bulls;
- hose and fittings;
- nose clips – these are compulsory for animals over 12 months;
- show cane;
- show products such as shampoo, shine, combs and brushes;
- neckropes – these are compulsory at all shows;
- leather show halters;
- fitting box (lockable) to store tack and equipment while at the show;
- grooming equipment and blowers;
- electrical leads and extension cords – there are never enough power outlets for everyone;
- curtains – some breed societies provide these and some studs make their own. These are to hang header cards from;
- header cards – name cards placed above the animal with its identification details. Some breed societies provide them but most exhibitors make their own in their stud colours.

Presentation

I consider showing is a marketing tool, so it is all about presentation. Remember you are there to promote your herd and your stud, so make sure you present it well. Keep your animal beds clean and tidy, keep your animals clean, make sure you go out in the ring with a clean set of clothes and polished boots and your animals are presented to the best of your ability.

Order of judging for stud stock

- On entering the ring the judge will ask for the animals to be paraded in a circle. In Australia this is always in a clockwise direction.

- Classes are usually judged with junior cattle first followed by senior cattle.
- Cattle normally enter the ring in order of youngest to oldest. Local shows are usually casual about this rule but royal shows should always follow this protocol.
- Judges normally view and assess each animal on an individual basis, perhaps asking you to lead them out individually and parade them in a straight line.
- Once their decision has been made, the judges will indicate for those to move forward in the order chosen to award the ribbons.
- The number of ribbons awarded will depend on the number of entries in a class. The schedule will stipulate the rules. Normally the first three places are recognised.
- The animals that win their class are then asked to come back into the ring for judging the age championship. Individual shows will decide how many junior and senior classes they have. Some shows also have calf championship classes.
- Once the judge has made the decision the animal that came second to the championship winner is asked to come back into the ring to have an opportunity to be judged for reserve champion.
- From the calf, junior and senior champions, a Grand Champion is chosen for both sexes.
- The Grand Champion from each breed is eligible to compete in the Interbreed Championship, where one animal from all competing breeds is selected as the supreme exhibit of the show.
- Some shows have a breeders' group, where you can exhibit three animals that you have bred and that carry your stud prefix. Usually both sexes are represented. The aim is to show the judge what you are trying to produce.
- Some shows have an exhibitors' group, where you can exhibit three animals that you own. Usually both sexes are represented.
- Sire's progeny classes usually comprise three animals, representing both sexes, all from the same sire. Occasionally dam's progeny classes are included, with exhibits of animals out of the same dam. This is sometimes more difficult to show.

Tips: Cattle often won't drink the water provided at a show or field day due to its different taste. One way to overcome this and to avoid dehydration is to add some molasses to the water. A tablespoon will suffice. You can get the cattle accustomed to this by adding some to their drinking water at home a few days prior to the event, so they are used to it beforehand. If this fails, don't panic – they will drink when they are thirsty enough. I also add electrolytes to the water for my show team to prevent any dehydration. Don't use communal water troughs – they are the easiest way to transfer viruses and diseases. Hand water all your animals from your own buckets.

This is what our site looks like on display at a royal show.

Show etiquette

- All bulls over 12 months must have a nose ring.
- You should not speak to the judge prior to judging even if they are your best friend.
- Remember to congratulate the winners.
- It is supposed to be an enjoyable experience, so make it that and don't get too anxious on the day.

10 reasons you know you enjoy showing cattle

1. Your family holiday is a trip to a cattle show.
2. You've used a blower to dry part of your clothes.
3. Your Christmas list is full of show supplies.
4. You spend more on hair products for your cows than for yourself.
5. You know more about the genetics and family of your cattle than your own.
6. Your calendar has a countdown to the next show.
7. Your room is a mess but your cattle shed is spotless.

8 Your showbox has functioned as a bed.
9 You learn to ignore getting wet at the washbay at 5 am.
10 Your teachers know where you're going when you ask for next week's homework in advance.

Showing led steers

In recent years I have had considerable success in the led steer arena. For me it was a natural progression from the stud ring, as I wanted to be able to demonstrate to clients that if they purchased one of my bulls for their commercial cows this is what they can produce. It is the perfect way to demonstrate your breed's commercial viability.

I can assure you that showing led steers is a walk in the park compared to showing stud stock! Once broken in, they are so much easier than stud stock as they are not cycling or worrying about their calf or fighting with the bull next door. Steers are quiet and gentle.

The most important factor with steers is that you are looking for carcass traits. Confirmation is not as important, as lots of led steers can't walk properly. Most purebred led steers are steers because they were not good enough to be a stud bull. So length, muscle, depth, eye muscle area, volume and fat coverage are the factors being judged.

The judge should be choosing the animal that he or she thinks would be the best carcass if it were sent to the abattoir and cut up on the day of judging. It annoys me if a judge says, 'I chose this steer as he will go onto make a great carcass.' That's not the point of the competition! The decision should be based on the most market-desirable animals on display in front of the judge. Remember that steer judging, like stud stock judging, is only one person's opinion on the day. You might win Grand Champion at one show and not even get placed at the next!

Avoid animals that are:

- cut up high in the flank;
- bad tempered and nervous – animals that get stressed often end up becoming 'dark cutters', which will disqualify them from a carcass competition as the stress has raised their pH level.

Desirable traits include:

- evenness of fat cover;
- a well-defined muscle pattern;
- good length to give a larger EMA (eye muscle area).

Another important factor with led steers is correct feeding. They need to fit within the nominated weight specification, so you will need to monitor their feed regime closely. You also want to be able to meet the fat specifications as required by each class. You will definitely need a set of scales if you want to be a serious contender in led steer competitions. I weigh my steers every week for six weeks prior to competitions.

- Led steer classes are based on weight, not breed, so all cattle of all breeds are exhibited together.
- Most class increments are in 50 kg at royal shows. Local shows can have a huge variation of weight, as often there are only two classes. There will be a minimum and maximum weight so, if you want to be competitive in led steer classes, you need to plan your steers' age and weight, normally 12 months in advance. I had to change my commercial herd to an autumn calving so I would have steers at the right age for the Royal Queensland Show.
- All led steers must have no visible signs of permanent incisors (teeth), which means they must be younger than about 18 months. Most cattle lose their milk teeth or baby teeth around 18 months, and some even earlier. They will be checked at the show and disqualified if they have any permanent teeth.
- Led steers require nose rings.

Hoof and Hook competitions

Royal shows in Australia and other big field days and expositions conduct Hoof and Hook competitions, in which steers and heifers are judged on the hoof by visual appraisal then sent to an abattoir to be judged on the hook. The hook section is scientifically measured, with the various sections being awarded on a point-scoring system. The carcass with the most points is the winner. The appeal is that there is no subjectivity in the final results.

The ultimate achievement for me has been to win Grand Champion Carcass twice at the Royal Queensland Show, against all breeds, all weights and over 250 other entries. The first year that I was a recipient I won both the lightweight carcass and the heavyweight carcass then went on to take out the Grand Champion carcass. Two years later I was successful again, winning the lightweight section with the highest points in the competition and so it was Grand Champion Carcass. In 2014 I also claimed the reserve Champion Lightweight carcass. The Royal Queensland Show is the most prestigious Hoof and Hook competition in Australia for steers, with huge competition and excellent incentives. What's more, it allows smaller breeders to enter with just one animal – many similar types of competitions run by field days and expos are aimed at commercial producers with a pen of three to six animals.

To me, the hook component of these competitions is far more important than the hoof component. The hook is measuring the yield and beef quality of the animal and, when we are in the beef industry, surely the aim is to produce quality beef. Very rarely does an animal chosen as the Grand Champion Led Steer then go on to win the Grand Champion Carcass.

What breeds work best?

When choosing animals for these competitions I always use 1st or 2nd-cross steers. I could not achieve the desired weight with a fullblood or purebred Lowline but possibly could with an Ausline. I like the hybrid vigour that crossbreeding provides. I am also selective in the types of breeds I use for my led steers, as certain breeds provide specific traits that I want in my steer and subsequent carcass. I prefer a European/British cross. The British breeds are easier to finish, providing the right amount of fat cover and giving softness to the animal, and they have excellent meat qualities. The European breeds provide muscle and growth but I find that too much European content makes them too coarse and too hard to finish.

I have had the most success with 50% Ausline, 25% Limousin and 25% Murray Grey. The Ausline gives the early maturity, softness and finishing ability. The Limousin gives that extra muscle and growth. The Murray Grey, another British breed, provides growth and finishing ability. I have also had a lot of success with Senepol/Ausline cross. The Senepol is a tropically adapted breed developed for its carcass qualities.

A steer that I bred was the Champion Lightweight Led Steer at the Royal Queensland Show in 2011.

Hoof section

The animal is first judged in the led steer class based on weight. At royal shows there are normally three lightweight classes, three medium-weight classes and three heavyweight classes. From within these sections there will be a Champion and a Reserve. The three champions will then compete for the Grand Champion Led Steer. Prize money at most royal shows is often very significant for these championships.

Hook section

As soon as the animals have been judged as led steers they are loaded onto a truck and sent to the nominated abattoir, where all the carcasses are judged by MSA (Meat Standards Australia) accredited graders using a uniform set of standards. Up to 10 different sections are measured. In each section a maximum amount of points can be awarded, and the carcass with the highest number of points is the winner.

An example of a specification sheet from the 2014 Royal Queensland Show is given on page 195. The judge has measured the fat depth on the ribs and on the rump (P8), the colour of the meat and the fat, the marbling (how much fat is visible in the meat), eye muscle area, meat yield (the percentage of useable meat), and how evenly the fat is distributed between the two measured sites.

14

Being a player in the beef industry

For many small acreage holders the idea of running a smaller type of beef cattle, not only to keep the grass down but also to make a small profit, is appealing. Before you venture down this track you need to evaluate your production capabilities and have very clear objectives. You need to have a plan and know which markets you want to supply, your breeding objective and how you will achieve it. Many small farmers have failed because the heart ruled the head when tough decisions had to be made.

- Do you want to breed your own animals or do you want buy in store steers and fatten them?
- Will you feed grass or grain?
- What breed of cattle will you use?
- What market are you aiming to supply and how will you access that market?
- What are the costs involved?

The ultimate accolade for me came in 2007 and again in 2009 when I won the Grand Champion Carcass at the Royal Queensland Show. The carcass of our home-grown steer was awarded the highest score of all 250 entries against all breeds. In 2012 I was a finalist in the Delicious Produce Awards, which is based on taste and judged by Australia's leading chefs. The same year I won a silver medal at the Sydney Royal Easter Show for my beef in the branded beef competition. This meant that small breed beef is now considered a serious contender within the beef industry.

This shows that if you have a plan to grow and diversify, it can be achieved. I started off with showing stud stock, then expanded to led steers and carcass competitions, which led to the development of my own branded beef. Alongside this production plan was an integrated strategic marketing plan.

I have always maintained that smaller cattle have finer-grained beef and that finer-grained beef is the most tender beef. I believe this finer grain comes from their shorter cell structure. Combine these excellent meat characteristics of smaller breeds with their early maturing patterns and you have a beef animal that fits supermarket and domestic trade absolutely perfectly. What's more, you can have a higher stocking rate with animals that are finished at a younger age.

Production type

First of all, you will need to evaluate the range of production options. This will depend on your acreage size, stocking rate and available pastures, length of growing season, rainfall, paddocks and fencing as well as your degree of knowledge. If you are not a full-time resident on your acreage, that may also affect your decision.

If your enterprise is to be profitable then you will need to work out your cost of production, measuring it in cents per kilogram and providing an indication of the outlay required to produce each kilogram of meat for beef. MLA has an excellent tool on its website which requires only 15 minutes for you to calculate your cost of production.

Breeding and finishing your own animals

Everyone loves the idea of breeding their own animals. Being able to choose your genetics and see the progeny mature can, like gardening, be very rewarding. But beware: it is not easy and while the idea sounds appealing there are more than a few problems you will meet along the way. But if the heart is not allowed to overrule the head, it can be a profitable enterprise. For systems like this to be profitable you have to be strict in your selection criteria and decisions. Non-productive females and bulls have to be culled immediately, not kept due to some personal attachment.

Do you have the space to run and segregate bulls, paddocks to separate weaners, adequate pasture to grow your weaners, someone to regularly check calving cows, strategies to deal with drought?

Buying and finishing steers

This system often suits those who don't reside full-time on their acreage or those who want the least maintenance. Steers are by far the easiest type of animal to rear and care for, they require the least commitment in time as they can run together without the worry of calving, weaning or bulls. Likewise, it is easier to sell if calves are not a consideration. Some country is considered poor fattening country, due to long cold winters or poor-quality native pastures, so it is best to check with other producers in your area before you make a decision. You have to be sure that you can make a profit from this system, and that will ultimately be determined by the selling price less the purchase and running costs of the stock. The profit margin can also be affected by seasonal conditions, which are out of your control, so if feed starts to decline you need to be able to act swiftly by either supplementary feeding or selling your stock. The problem is that lots of other producers are in the same position – the markets become flooded with producers trying to offload their stock in dry conditions.

My cousin engages in this practice on his 5000 acres, which is mainly cropping but in the back country he can run cattle. He buys in weaner steers and runs them for another 18 months. They get very little attention with regard to animal husbandry. By the time he takes into consideration cartage fees, saleyard fees etc. his profit is about $100 per head.

Backgrounding

This refers to contract growing, where an outside party such as a feedlot or stud producer owns the animals and pays you for the weight gained on their animals. With this system you are often not expected to fatten the animals, just grow them out. This takes the pressure off feed availability so it may be a better option than grass finishing. You are normally responsible for health treatments and are paid only on the basis of live weight gained – any death or losses are deducted from the total.

Production system

Another factor that needs to be taken into consideration is how you intend to finish your animals: on grass, on grain assist or lot feed them on grain. The term 'finish' means to have them market-ready with the right amount of fat coverage and weight. Obviously grass is the most cost-effective and easiest method to pursue but you need to make sure you will have adequate pasture at the right time. Grain assist means that the cattle have access to pasture but are supplemented with grain. Lot feeding means a diet of 100% grain and some roughage, no pasture and housing in confined areas. There are a number of specific guaranteed production methods that can require certification.

In my opinion, the taste of grass-fed far outweighs that of grain-feed beef. Cattle were not designed to be locked up in feedlots and force-fed grain continually for up to 300 days. They can't metabolise it and the resultant product tastes nothing like beef should. They have no access to grass, have no room to move, and are injected with growth promotants and chemicals, which then build up in the fat. These livestock often end up with a number of illnesses. The resultant product has no flavour and most of the time is cryovacuumed, which I believe also destroys the texture of the beef and makes it taste like paste.

A study released by the University of Rochester Medical Center in New York held that a high consumption of beef by pregnant women may have contributed to low sperm counts in their sons. It is interesting to note that hormones have been banned in beef production in the EU since 1988. It's a shame that Australia does not do the same, but a number of supermarkets are trying to source beef that is hormone-free.

Meat flavour is also directly correlated to age. The older the animal, the more flavour it has. In Europe, much older animals are slaughtered for their distinctive flavour. Meat quality mainly depends on stress levels. Highly nervous or stressed animals produce tough beef. Meat tenderness depends on a rising plane of nutrition, good-quality feed and genetics. Certain DNA markers can test for the tenderness, marbling and feed efficiency genes, which are hereditary and can be passed on to progeny.

To be able to produce grass-fed beef you may have to run fewer stock so that you have enough pasture to last year-round. Not all producers are able to finish cattle off on grass alone. Sometimes they have to grain assist, which involves giving cattle supplementary grain in addition to pasture in times of poor pasture growth due to lack of rain or severe temperatures. These costs need to be taken into consideration when planning your beef enterprises as they can dramatically affect your bottom line.

Certified pasture-fed (PCAS)

The recent introduction of the Pasturefed Cattle Assurance System (PCAS) is a program that enables grass-fed cattle producers to market their cattle as grass-fed with certification. Cattle are required to have access to grass all their life (no grain or grain by-product whatsoever), are fully traceable all their life, are free from hormone growth promotants (HGP) and are free from antibiotics. For many producers this is a much more achievable accreditation system and it is already providing significant premiums for them. Certification requirements can be found at certifiedpasturefed.com.au.

Organic

Organic grass-fed beef is the most expensive beef in the US and soon will be worldwide. In 2014 the organic industry was worth $1.72 billion to the Australian economy and it is growing at 15% per annum.[7] The health benefits, combined with the flavour, make organic grass-fed beef one of the fastest growing commodities in the US and in Australia. Organic production requires a paradigm shift in a producer's way of thinking, where high productivity does not necessarily mean high profits. More is not necessarily better. It also requires a producer to think differently about how to deal with parasites, fertilisers and chemicals; some producers find they cannot be totally organic. For example, our property is in a tick area and because we show stud cattle they have to be treated with a government-approved chemical before they will be approved to enter a show ground. This means that we cannot be certified as organic producers if we want to continue showing stud stock. Also our breed of cattle, being British, does require tick treatments in the summer and the only products that are effective are not classified as organic. We would either have to change our breed to one that had a higher *Bos indicus* content for parasite resistance, move our enterprise to a tick-free area or give up showing stud stock. None of these options is appealing at this stage. Having said that, however, I have two clients who have become organic with the Ausline breed and they both reside in a tick area. Their methods have almost eliminated the ticks from their property. There are a number of different organisations with which you can affiliate to become accredited as organic, so do your research first. Certification is usually a three-year process but proposals are under way to reduce this to one year to include acknowledgement of prior practices. Accreditation involves audits and fees. The National Standards are available at agriculture.gov.au.

Biodynamic

This is an agricultural system that requires specific additional requirements to an organic system. It is based on the application of preparations developed by Rudolf Steiner. The method uses biological–dynamic practices to activate the life of soil and plants. Certification is normally a longer process than organic accreditation. The standards are available at Demeter.org.au.

Range-fed

We market our beef as range-fed. This means that it has no growth promotants, is as organic as possible by engaging in rotational paddock systems to reduce parasite infestation and subsequent drenching and, when seasons allow, the cattle are fed entirely on grass. We use organic fertilisers, don't spray our crops with chemicals and only worm cattle less than 12 months of age. In times of pasture shortage we finish our cattle by grain assisting them for 50–70 days, during which they still have access to grass. My meat has been tested by an accredited laboratory to see how the omega 3:omega 6 ratio would alter, if it did so at all, by finishing my beef on 50 days of grain. Because the cattle were raised entirely on grass for their first 16 months there was no change to the ratio.

Certified grain-fed

Cattle slaughtered and processed as grain-fed must be sourced from an accredited feedlot and audited by Aus-Meat. All animals must be fed a ration of not less than

> Did you know a cow emits about 600 L of methane gas every day? This is enough to fill up 40 party balloons!

10 MJ/kg of dry matter. There are a number of categories but adult cattle (Symbol GF) require a minimum of 100 days with a maximum of six permanent teeth and a fat depth of 7 mm. Young cattle (Symbol GFYG) require a minimum of 60 days for females and 70 days for males. They require none to two permanent teeth and a fat depth of 5 mm.[8]

What breed is the best?

Beef is not breed-specific and by that we mean that different breeds do not have a specific flavour. Once slaughtered and cooked, no one is able to tell the breed of the beef you are eating. Diet certainly affects the flavour and taste of the beef, but the breed does not.

Certain breeds have characteristics and qualities that are recognisable in the beef, such as Wagyu and its high marbling. The market that you decide to supply will determine what breed you use. European breeds are known for their higher muscling and British breeds for their softness and their finishing, which is the ability to lay down fat. *Bos indicus* cattle are known for their parasite and heat resistance traits as well as their foraging abilities, and as such as considered more hardy and able to survive in tougher conditions. You need to recognise these different breed strengths and weaknesses such as muscling and ability to lay down fat, growth, fertility, temperament and maternal traits.

The market you choose will determine what frame size you are looking for in your breeding program and therefore which breeds are most suitable. Once you have established the market you are aiming to supply, you need to determine what breed you will use to meet its demands.

Many producers who raise cattle for the beef market use a crossbreeding program that allows them to increase the productivity of their herd. Crossbreeding allows for hybrid vigour (heterosis). Hybrid vigour is the increased performance associated with the combination of parents from two unrelated breeds. The increased performance traits are usually most noticeable in growth, fertility and survival rates. By using the right combination, increased growth rates of up to 40% have been achieved. True hybrid vigour occurs when the crossbred animal is a terminal cross, meaning that both parents must be purebred. Hybrid vigour is usually most evident when the breeds joined are genetically the least related, such as a *Bos indicus*/*Bos taurus* cross.

Smaller breeds of cattle offer a definite advantage. Their genetics are based on high 200-day growth, moderate 400-day growth and low 600-day growth. With these growth patterns, producers can enter the exclusive end of the domestic market, by producing steers that can fatten and marble off grass or with 60–90 days of grain finishing.

The commercial advantage begins to emerge at about 12–14 months when the animal starts to mature and begins putting its energy into developing muscle and fat rather than bone growth and body form. A commercial operator will be producing steers of 300–350 kg that are already starting to lay down fat. Many breeders of other traditional breeds argue that their cattle will be 50 kg heavier at this age. While this may be true, the difference lies in the cost of getting those animals to an eating standard, i.e. the cost of finishing that animal, which is the major cost to meat producers. With the early maturity of a small breed, producers can sell their stock without the need to supplementary feed. Time and time again we see producers trying to fatten their large-framed steers that continue to grow rather than produce fat. Small early-maturing breeds help 'downturn' the high growth curve of regular cattle so that they can mature earlier and express fattening ability at an earlier age.

On top of this, the smaller breeds are able to produce an exceptionally high percentage of eye muscle area (EMA) for their size. This is an important statistic as EMA or eye fillet (also referred to as primal cuts) commands a much higher price. If you consider the photo on page 37 the Dexter bull weighed 540 kg with an EMA of 101 cm (18.7%) compared to the Red Angus bull whose weight was 970 kg with an EMA of 134 (13.8%). The Dexter is producing more EMA as a percentage of his carcass than the Red Angus bull.

Q. My in-laws have just purchased approx. 100 acres (with a house) in the area and we are keen to run cattle for beef. Being typical ignorant city slickers, we are unsure as to what breed of cattle to choose.

A. Small breeds are very suitable as beef animals. The bulls can be used over a range of different breeds to produce a commercially viable beef carcass. You would need to choose a breed that is suited to your environment. Tropical cattle don't survive too well in the snowfields and many British breeds don't thrive so well in the tropics of northern Australia. Having said this, however, I have sold Lowline to the Northern Territory and to Tully in far north Queensland. They survive because they are well cared for and their owners make sure they have adequate shade in the warmer months.

Market options

You will need to know your objectives with regard to what markets you are aiming for. Each has its own market specifications and there are a number of factors that will affect your price. They include weight, age, sex, muscle pattern

Small breeds can easily produce show-winning commercial quality steers. A number of small breeds consistently do well in national hoof and hook carcass competitions. At the Royal Queensland Show in 2014 three Ausline steers were placed first, fourth and sixth in the same class. The steer on the left went on to win Reserve Champion Lightweight Carcass Steer. Left: 16-month-old steer 25% Ausline, 37.5% Murray Grey and 37.5% Limousin weighing 362 kg. Middle: 16-month-old Ausline (50%) and Belmont Red (50%), weighing 362 kg. Right: 12-month-old Lowline (50%), Limousin (25%) and Murray Grey (25%), weighing 360 kg.

and fat coverage. On top of this you will need to assess the current market price, which is influenced by supply and demand and current seasonal conditions. I find it curious that Australian cattle prices are 50% lower than cattle prices in the US, yet Australian consumers pay 20% more for beef at retail than consumers in the US.

A plan is essential before you start. Your decision on which market segment is the intended destination of your product, will be determined by a number of factors:

- size of property and scale of production;
- what breed of cattle you are carrying;
- carrying capacity of the land;
- rainfall;
- pasture;
- market versatility, i.e. can you move to another market in time of price fluctuations or drought;
- ability and continuity to supply;
- distance to your market.

Weight

Your breed and pasture conditions will be the most limiting factors in choosing the weight you aim to produce. Normally markets use the following categories:

- lightweight, also known as domestic butcher trade:
  - weight range 325–400 kg;
  - optimum P8 fat depth is 9–10 mm;
- medium and heavyweight, also known as supermarket and domestic trade:
  - weight range 400–500 kg;
  - optimum P8 fat depth is 10–11 mm;
- heavyweight, also known as the Japanese chiller trade:
  - weight range 500–710 kg;
  - optimum P8 fat depth is 11–12 mm.

MLA has an excellent online BeefSpecs calculator, which is a tool to assist in meeting market specifications. http://beefspecs.agriculture.nsw.gov.au.

How to sell livestock

Another consideration will be how and where you want to sell your stock. All options have disadvantages and advantages and there is no single best selling method. The size of the mob, transport costs, the type of stock, importance of carcass feedback and net return will affect your choice of a selling method.

Saleyards

This is the most common option for producers. The cattle are either sold on cents per kg of live weight (c/kg) if the saleyards have scales, or dollars per head ($/hd) if they don't.

Advantages:

- all types of stock can be sold and in any number;
- buyer assumes all risk once the hammer falls;
- competition from a number of buyers from different sources;
- sale is immediate;
- market price is guaranteed.

Disadvantages:

- additional costs include transport, yard fees, levy and commission;
- price depends on the number of buyers present on the day;
- heifers usually command a lower price;
- no carcass feedback.

On the hook sale

This is where the abattoir buys your animal directly. The cattle are either sold on cents per kg of dressed weight (c/kg dressed weight) or priced individually according to carcass measurement. Price will depend if the animal is traded on a hot weight or a cold weight.

Advantages:

- carcass feedback is provided;
- reduced stress due to less handling and transport;
- price is paid on quality and meeting market specifications;
- premium paid for better quality carcasses.

Disadvantages:

- lack of competition of buyers;
- price will be reduced for excess fat, weight and age;
- no bargaining power;
- not all types of stock maybe accepted.

Paddock sale

These type of sales can be negotiated with or without an agent, for all types of cattle. Many producers deal directly with a retailer, such as a supermarket chain or butcher, and cattle are paid as dollars per head or cents per kg live weight. Building alliances and long-term relationships can be a very cost-effective way of selling your product.

Advantages:

- direct transport to abattoir, reducing stress;
- ongoing relationships with buyer can be established;
- feedback can be provided;
- no agents are necessary;
- prices can be negotiated and guaranteed.

Disadvantages:

- lack of competition;
- if a price is predetermined then it may not always be the best;
- sometimes difficult for those with small numbers.

AuctionsPlus

This is an online system where animals' details are entered into a computer and they are sold from the farm. AuctionsPlus allows commodity transaction, reserve price-setting and legal change of ownership without the seller, buyer or product having to meet physically at the time of the sale. This is live online in real time, but is interfaced with a physical auction that is being controlled by an auctioneer. The cattle are assessed independently prior to the sale. Cattle can be sold on any basis, which is predetermined. Bids are taken on computer or by telephone.

Advantages:

- the purchaser pays transport costs;
- transit insurance is provided by AuctionsPlus;
- payment is guaranteed;
- competition is national;
- suits producers who are geographically isolated.

Disadvantages:

- buyers are unable to visually appraise animals;
- adjustment to selling via this method.

Producer alliances

For smaller producers, an alliance is a great way to supply a market. This is particularly relevant for continuity of supply, as often those on smaller acreage simply cannot produce enough to supply a market. There is a wide range of producer alliances, some breed-specific (e.g. Angus beef) or production-specific (e.g. organic beef). A group can accomplish more than an individual producer can.

Advantages:

- increase in scale;
- continuity of supply;
- ability to act quickly on consumer feedback.

Disadvantages:

- group dynamics;
- expensive to set up;
- criteria have to be adhered to in order to be successful;
- producer commitment.

Farmers' markets

An increasing trend is for producers to attend farmers' markets to sell their product. While the idea may be romantic, it may not be that easy. From my experience, the hardest part is getting a butcher to cut the meat to your specifications. Farmers' markets are gourmet and, not to be derogatory, not all country butchers are gourmet-oriented. They can do the basic cuts, so you need to develop a relationship so you can present your cuts the way your customer desires. Be warned: this will be a steep learning curve and it may take one or two attempts to find the right person. It's very expensive to kill and process animals on-farm if you intend to sell the meat. It's also illegal, unless you have a purpose-built facility, accreditations, audits and a full-time meat inspector. Even if you want to get your butcher to cut up your meat then pack it yourself on-farm, you need accreditation. Likewise, your transport vehicle needs to be accredited and audited as food-safe.

Advantages:

- higher returns;
- established relationships with customers;
- no middle-man expenses;
- flexibility to modify your product as demand requires;
- value-add to by making other products.

Disadvantages:

- incredible amount of bureaucracy – if you want to cook samples, you will need a licence from the relevant council;
- expenses including fuel, market stall fee, accreditation and audit fees, butchering fees, abattoir fees, insurance and registration, packaging and supplies, marketing, electricity and cold storage, printing and labels;
- dependent upon the weather … bad weather means no customers;
- regular weekend commitments.

Home kills

One of the benefits of home-killed beef is that the animals suffer no stress prior to slaughter, which makes for more tender beef. They are happily standing there eating grass then suddenly, with one bullet, they are dead. They did not have to be loaded onto a truck, driven to the abattoir, unloaded into unfamiliar yards then ushered up a race with electric prodders.

Advantages:

- no stress;
- no transport or agents' costs;
- performed where and when it suits you;
- carcass is cut up to your specifications;
- inexpensive process.

Disadvantages

- need space to store a whole beast;
- not legally able to sell any meat;
- have to be able to dispose of waste;
- time-consuming.

Costs

When considering your future beef operation, the physical features of your farm will limit your capacity. So you may need to consider the costs involved in upgrading your facilities, such as:

- degree of pasture improvement;
- yard facilities;
- paddocks and fencing;
- watering points.

In addition you will need to factor in other costs, such as:

- supplementary feed;
- drenches;
- veterinary costs;
- labour;
- repairs;
- marketing;
- selling fees, i.e. transport costs, agents' commission, killing fees.

If you are considering running cattle in the hope of covering your expenses as well as the interest on the borrowed capital, then do your sums thoroughly. You will need a very slick operation to do all that and make a profit as well.

Terminology

If you are going to produce beef, there are some basic terms that you will need to know and understand. There are a number of different factors that are measured in an animal's carcass when it is graded.

Age: This is an important factor in meat eating quality. Cattle are aged by their teeth, otherwise known as dentition (Table 14.1). This refers to the number of permanent teeth in the lower jaw, which gradually replace their milk teeth. An animal's breed and feed can affect when it gains its next set of permanent teeth.

Dressing percentage: Also referred to as the animal's carcass weight. The formula is:

$$\text{Dressing \%} = \frac{\text{carcass weight}}{\text{live weight}} \times 100$$

Bos indicus cattle normally dress higher than British breeds due to their lighter gut contents. European cattle normally dress higher than British due to heavier muscling.

Table 14.1: Relation of teeth to age of cattle

Permanent teeth	Approximate age
None (milk teeth)	<20 months
Two	20–30 months
Four	24–36 months
Six	30–42 months
Eight	>36 months

Eye muscle area: Referred to as the EMA. This is the measurement of the eye fillet in square centimetres. The larger the EMA, the more desirable the carcass.

Fat colour: Fat is graded on its colour. Colour has no relationship with eating quality but it does play a role in the presentation of the meat. Consumers prefer white fat, so white is given preference. Grain-fed cattle produce white fat, while yellow fat is often seen in older and grass-finished animals. Some breeds of cattle, such as Jersey, also produce yellow fat.

This is a personal beef of mine, pun intended. If I enter a grass-fed animal in a carcass competition and it has yellow fat, as nature intended, I am penalised for its colour. How ridiculous is this? Consumers have been conditioned to think that yellow fat is undesirable.

Fat depth: This is also referred to the finish of the beast. Selling animals with the right amount of fat coverage is very important. Those that are too fat are heavily penalised, as too much fat means wastage and this has a direct correlation with how much saleable meat the butcher can obtain from the carcass. Animals that are too lean tend to produce poorer eating quality beef.

Fat is measured in two places: the P8 (on the rump) and over the rib eye (usually taken between the 12/13th ribs). The fat measurement on the P8 is normally 20% higher than that taken from the rib.

Hot carcass weight: This is the term used in the meat industry. It is the weight of the animal once it is killed and the hide and certain other parts have been removed. The removed parts include the feet, head, intestines, udder, testes or penis, channel fat, kidneys, intrathoracic and abdomen fat. Carcasses also lose up to 3% of their weight when they are chilled.

Live weight: This is the weight that is used on the farm and in the saleyards. It is the total live weight of the animal.

Marbling: This is the intramuscular fat within the meat. It affects not only the flavour of the meat but the eating quality. It consists of fine flecks of fat within the muscle. This intramuscular fat is different from the subcutaneous fat that is also measured. Highly marbled beef is very sought-after by the Japanese and restaurants throughout the world. The longer an animal is fed grain, the higher the levels of marbling that can be obtained. Certain breeds such as Wagyu and Jersey produce more highly marbled beef, a trait that is considered to be moderately heritable.

Meat colour: Bright cherry-coloured meat is preferred by consumers. Darker meat often is often associated with a high pH and reduced shelf life. The eye muscle area is used to determine the colour of a carcass.

MSA: The Meat Standards Australia grading model predicts the eating quality of 40 muscles by six different cooking methods. MSA involves all sectors of the supply chain from paddock to plate. If carcasses fail to meet MSA minimum requirements, they are below the benchmark standard for high-quality table beef. You need to be an accredited producer and use an MSA-accredited abattoir to have your carcasses graded under this system.

Class 2B **Led Steer Carcase - 351kg to 375kg**

	Exhibitor	Weights: Animal Live Wt.	Weights: Animal Dressed Wt.	Weights: Dressed %	Fat Depth P8: Depth (mm)	Fat Depth P8: Points (Max 15)	Fat Depth Ribs: Depth (mm)	Fat Depth Ribs: Points (Max 15)	Fat Colour: Score	Fat Colour: Points (Max 10)	Meat Colour: Score	Meat Colour: Points (Max 15)	Marbling: Score	Marbling: Points (Max 5)	Meat Yield: %	Meat Yield: Points Max 10)	E.M.A: Area (cm2)	E.M.A: Points (Max 15)	Fat Dist.: Points (Max 15)	pH	Carcass Total: Points Max(100)	Hoof Section: Points Max (50)	Overall Total: Points (Max 150)
208	Hayes, Margo	362	203.4	56.2%	10	14	4	14	0	10	1b	15	1	5	65.7	10	79	15	9	5.64	92	40	132
214	Cowgirl Cattle Company	368	224.8	61.1%	6	12	5	15	0	10	1c	14	0	4	64.79	10	86	15	10	5.54	90	46	136
207	Steele Rudd Murray Greys	374	222.8	59.6%	7	14	5	15	0	10	1c	14	2	4	63.61	10	75	13	9	5.66	89	40	129
210	Hayes, Margo	360	202	56.1%	10	14	5	15	0	10	1c	14	1	5	63.19	10	64	10	10	5.57	88	40	128
203	Tenterfield High School	374	218.8	58.5%	5	8	6	14	0	10	1c	14	1	5	64.12	10	83	15	12	5.39	88	38	126
209	Hayes, Margo	362	201	55.5%	8	15	6	14	0	10	1c	14	1	5	62.57	9	64	10	10	5.48	87	40	127
219	K.B.V. Simmental Stud - M Rowlands & S Lean	366	207.2	56.6%	10	14	6	14	0	10	1c	14	1	5	63.31	10	72	12	7	5.36	86	38	124
201	St. Johns College Dubbo	372	227.6	61.2%	6	12	4	14	0	10	1c	14	0	4	66.5	10	97	15	4	5.41	83	40	123
202	Holy Trinity School &, Mr D Strelitz	354	217.8	61.5%	5	8	4	14	0	10	1c	14	0	4	66.35	10	91	15	7	5.39	82	43	125
215	Benson, Brian and Esther	368	205.6	55.9%	5	8	3	12	0	10	1c	14	0	4	67.61	10	90	15	7	5.59	80	38	118
218	L J & L R Lee	366	207.2	56.6%	10	14	7	12	1	8	2	11	1	5	61.92	8	66	10	12	5.45	80	43.5	123.5
204	Dubbo College Senior Campus	366	206.2	56.3%	7	14	5	15	0	10	3	5	0	4	64.13	10	73	12	9	5.34	79	40	119
211	M F & N J Nicholls	362	204.2	56.4%	5	8	3	12	1	8	1c	14	1	5	66.56	10	81	15	7	5.45	79	44	123
217	Hunday Pastrol Co.	374	212.6	56.8%	5	8	5	15	0	10	2	11	1	5	63.59	10	71	12	7	5.39	78	44.5	122.5
212	Hillsborough Limousins	354	223	63.0%	5	8	3	12	0	10	1c	14	0	4	66.45	10	89	15	4	5.67	77	38	115
205	G A & L J Ball	368	199.2	54.1%	13	8	8	10	0	10	2	11	0	4	61.83	8	68	12	12	5.43	75	45	120
206	O'Dwyer, Mark	354	208	58.8%	4	4	3	12	1	8	1c	14	1	5	65.21	10	72	12	6	5.53	71	45.5	116.5
216	Livingstone, Jason & Megan	368	235	63.9%	2	0	2	0	0	10	1b	15	0	4	69.34	10	116	15	6	5.44	60	38	98

Carcass results from 2014 Royal Queensland Show Hoof and Hook competition.

Muscle pH: This is a measure of acidity and is a significant indicator of meat quality. After slaughter, lactic acid is formed naturally in the meat tissue, causing muscle pH levels to drop from around 7 (neutral) to an optimal range of 5.4–5.7. Excessive stress before slaughter can cause lactic acid to be produced as the carcass chills. This causes high pH, which can give the meat a dark colour and sticky texture. This makes the meat unattractive and reduces its shelf life.

Yield: This is the proportion of a beef carcass, excluding fat and bone, that is saleable as meat. It is a very important percentage to butchers as this is how they earn their income. Aus-Meat claims that the average carcass yields 68.7% meat,

Hoof and Hook competition. Left: This Lowline cross was fourth in his class on the hoof (338 kg). Right: It went on to win the Grand Champion Carcass (192 kg).

12% fat and 19.3% bone. Smaller breeds have been reported by some butchers to yield as high as 75% due to smaller bone content.

Basic cuts of a beef animal

The diagram shows what cuts of meat are obtained from what part of the beast.

The following list shows how the basic cuts can be marketed in the butcher's shop and what they are used for.

1 Shin: shin beef or gravy beef, osso bucco.
2 Round: round steak, beef strips or minute steak.
3 Silverside: corned, diced or silverside roast.
4 Topside (opposite side of silverside): topside roast, mince or steak.
5 Rump: steaks or roasts.
6 Sirloin: steaks or roasts.
7 Fillet: eye fillet whole or steaks, butt fillet.
8 T-bone: T-bone steaks, porterhouse steak, New York cut (no bone), roast.
9 Thick flank: skirt.
10 Set of ribs: standing rib roast, scotch fillet, rib fillet, cube roll, rib cutlet.
11 Rib-eye.
12 Chuck: diced or minced.

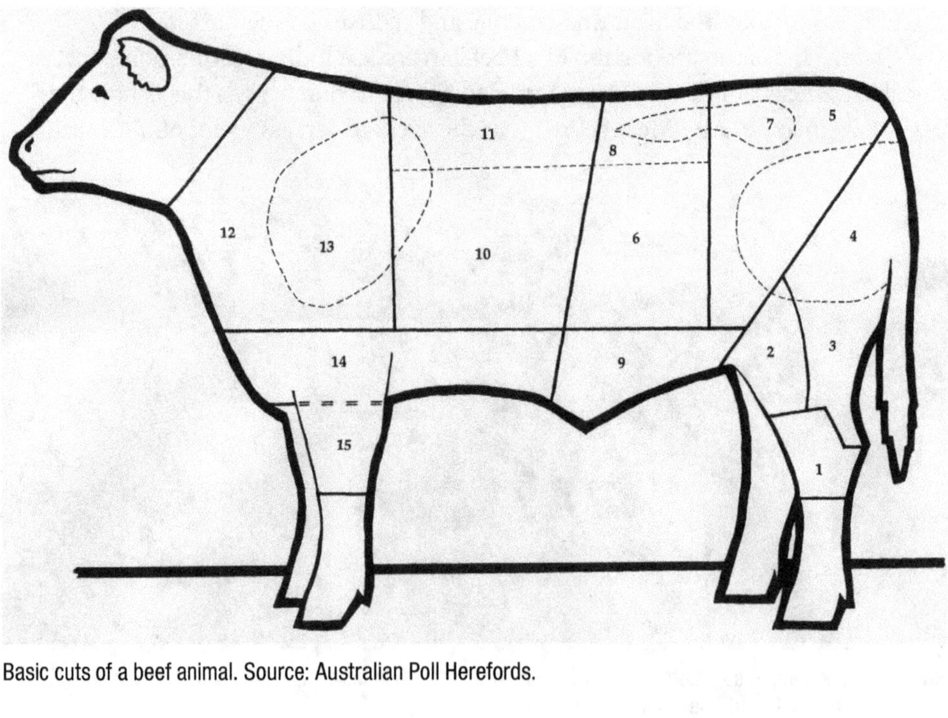

Basic cuts of a beef animal. Source: Australian Poll Herefords.

13 Blade: oyster blade, steak, roast, diced, minced.
14 Brisket: roast, corned, diced.
15 Shin.

MLA has launched a smartphone app 'Meat Cuts' to help consumers improve their knowledge of meat cuts and cooking methods.

15

Marketing

Marketing is an integral part of any business and one that is a continuous cycle. It involves product development, pricing, packaging, distribution and communication of the marketing message, including the process of making the sale. Don't confuse marketing with sales – selling is not a substitute for marketing. You need all parts to combine.

> **Marketing:** the whole is greater than the parts.

- If you decide to hold an auction in town and you paint a sign saying 'Stud cattle auction at the showground this Saturday' that's *advertising*.
- If you ride a bull down the main street of town with a banner, that's *promotion*.
- If the bull walks into the main bar and meets the mayor, that's *publicity*.
- And if you get the mayor to laugh about it, that's *public relations*.
- Finally, if you get good attendance after you explain what quality animals will be offered and people ultimately spend a lot of money at the auction, that's *sales*.

Often a stud herd is started and only two years down the track is the idea of marketing addressed. Some prior planning will make the job much easier. To be a successful marketer you have to think outside the box – think differently, use strategies and ideas that others haven't considered using before. The principles of marketing are the same for any business, they just have to be adapted to your particular industry.

Before you get started

Know why you want to market

There is a wide range of reasons but perhaps the two most important are to win business and to reinforce your product, getting your brand name out there and known.

Know your customer and remember the customer is always right

You need to know some basic information about who will be buying your stock, such as:

- Geography: what area do they reside in?
- Demographics: age and sex.
- Psychographics: lifestyle characteristics, behavioural patterns, beliefs and values. Don't waste time and money advertising your stock in a magazine targeted at those who reside in the inner city.
- Geodemographics: cluster marketing or lifestyle marketing. This is based on the old adage that birds of a feather flock together, i.e. like-minded people live in the same area and have similar backgrounds and consuming patterns. It would be unwise to start a small breeds stud in an area where the neighbours all have 10 000 acre properties.
- Spend time with your customers and listen to them. Find out why they want cattle and what they hope to achieve.
- Find out how they heard about your stud. This is good way to gauge if certain publications or advertising features are actually working.

  Small breed customers are typically those who own small acreage.

- Both partners work, i.e. dual income.
- There is a high percentage of female involvement.
- They have no prior involvement with cattle so want small quiet animals.
- They require lots of after-sales service.

Know your market and your product

- Buying decisions are rarely about price alone, but they are about value. Whatever price you charge for your product (animals, semen, embryos, beef), that price must accurately reflect the way your customer values what you are offering for sale.
- Customers assess value on a number of factors including quality, features, convenience, reliability, expertise, support and price.
- If you are going to be a leader in your field your customers will expect you to be superior in some aspect, such as having the best reputation, the best product or the best after-sales service. For example, BMW = best performance, Rolex = best quality, Rolls Royce = best luxury.

- Find out who your competitors are and why people buy from them. What is the competition doing that is different from you? Competition will include other breeders, other small beef breeds, other animals for small acreages, i.e. alpacas, mini horses, goats, beef industry.
- You want to increase your market share but beware, don't try to 'buy' market share through price reductions. In 2013 McDonald's spent more than $988 million on advertising and Coca-Cola spent more than $3.3 billion but neither brand, in fact no one brand in the world, has 100% brand awareness let alone market share. 25% is considered a dominant market position.
- Selling is the easy part if you have confidence in your product and you know it inside out and back to front. You just need to get the message out there.

Set your budget

Work out how much money you want to spend on an annual basis and what you hope to include in this budget. You need to consider what will be needed at shows and field days if you intend to exhibit at them. We include all our showing expenses as part of our marketing budget. Other items that may need to be considered include:

- brochures and brochure stands;
- business cards;
- banners/marquees/flags;
- stud signs;
- website.

Once you are established and start making sales, you will want to know how much it costs to get your customer. Here is how you can estimate what it costs you to bring in a new customer.

1 Add up the total of last year's marketing program.
2 Delete costs that were aimed at existing customers, such as newsletters etc.
3 Estimate how many new customers you think you acquired.
4 Divide the number of new customers into this amount.

This will give you an eye-opening indication of the value of every new customer. Make sure everyone involved in your business knows this value as it will certainly make them aware of what a valuable commodity they are dealing with each time they have customer contact.

Develop your strategy and make yourself stand out from your competitors

Image

First impressions are everything. Approach your stud. Drive up to it yourself. Does it look presentable? Make it attractive and have adequate visible signage.

At Vitulus, presentation and attention to detail are extremely important whenever the stud is showcased as these reassure our clients about the quality of our product. You wouldn't eat at a restaurant where the bins are overflowing, the tablecloths are dirty and the floor is filthy. You would be worried about the hygiene level, due to the bad image, and the fact that it may be reflected in the food they serve. Similarly, if people visit your stud and the fences are falling down, you don't know which cows were joined to which bull or when they are due to calve, then you're not offering a reassuring environment.

Service

Making the customer your number one priority allows you to offer a unique and differentiated service. As stated, many new clients will have very little prior knowledge or equipment so be prepared to help for as long as they need. If someone wants to buy show quality cattle then make sure you sell them show quality cattle. They will be promoting your brand for you in the showring, so make sure it is stock that you are proud of.

Innovative marketing strategies

Strategies that are different from those normally used within the beef industry, including an extensive website, presence on social media, on-farm field days, YouTube uploads, newsletters, and comprehensive information kits. The list is only limited by your imagination.

Value adding to the product

When a client purchases from our stud they get the whole package which can include a lot of extra services such as comprehensive after-sales support, delivery of product, product guarantee, expert advice, DNA verification, GeneStar® testing, semen packages etc. They can attend education days, buy the book or buy our beef.

Success is reinforced

Success in the showring, carcass competitions and branded beef competitions are advertised and displayed as clients like the reassurance of purchasing from a successful stud. It allows them to continue and share in that success and they like to be associated with success. Acknowledge your clients in your advertisements if they win with your stock. Remember they are promoting your brand for you.

Making the most of printed material

The choice and use of type is crucial, even in the letters you send. Make sure any literature you produce is easy to read and communicates your message clearly. Otherwise the money you spend on printing and distribution can easily be wasted.

While there are many fonts available, they fall broadly into two family categories, being serif and sans-serif. Serifs are the little 'feet' at the bottom/top of the descenders/risers. The words you are reading right now are in a serif font. Generally speaking, body copy should be set in serif font, as it makes it easy to flow quickly through the text. Nearly every newspaper or book you pick up uses a serif font such as Times or Courier.

Sans-serif fonts, like this one, do not flow as well. They should be used in headlines or other areas where the reader has to digest only a small amount of information.

ONE THING YOU SHOULD NEVER DO IS SET BOLD COPY IN CAPITALS. WHEN WE READ, WE ACTUALLY RECOGNISE THE SHAPES OF WORDS, RATHER THAN INDIVIDUAL LETTERS. IF YOU WRITE IN CAPITALS, THERE ARE NO RISERS OR DESCENDERS AND THE WORDS ALL BECOME RECTANGULAR BLOCKS. THIS IS VERY DIFFICULT TO READ, ISN'T IT?

Strategies to win new customers

Branding

Your name is your brand. Brand recognition means that your name is displayed whenever possible. Make use of every opportunity and display it as prominently as possible on staff uniforms, car logos, stickers, fitting boxes and caps. You build your brand as a result of the impressions made by encounters with your name, logo, marketing messages and everything else that people see and hear about your business. Every time a person goes to your stud, lands on your website, glances at an advertisement or meets an employee, they form an impression that leads to a mindset about your business. Everything about Vitulus is red – our cars, our uniform, our stock crate etc. So much so that if I go to a show and don't wear my red jeans, I get many comments as to why I am not wearing red!

Portray consistency: develop a consistent look, keep a consistent tone in communications and project a consistent level of quality that consumers can identify and relate to your name.

Develop a tag line

A tag line is also known as a slogan. 'Small Cattle with a BIG steak in the Future' was our initial line, then developed to 'Commercially viable smaller Beef'. It helps consumers link our name to our brand message. Make your tag line part of your logo and it will reinforce your logo.

Community involvement

Get involved with local shows and local schools. It helps get your breed's name out there in a very competitive environment and, what's more, it's very rewarding. We have had a lot of involvement with schools and students and we love their enthusiasm and energy.

Establish media relationships

Develop good relationships with relevant media bodies. Provide written information and photos if possible, as they are always looking for new and interesting stories. Remember you need to provide a different angle to make your story newsworthy. Don't just give them a story about you and your stud. Give them a story about a set of twins, a particular award won or the local school's involvement. Remember, it has to be something different.

Field days

For seven years we ran an annual on-farm field day. It was very successful and allowed existing and potential clients to see what we are all about, meet the staff and see what the cattle look like. It was also an educational event. It provided a greater knowledge and understanding of the stud's values, goals and product, which is sometimes difficult to convey through printed material. We then expanded this to the Small Farming Expo which is held at different locations each year and opened up even more marketing opportunities.

Market research

Do detailed market research to allow for identification and development of new markets such as famers' markets, dairy industry and commercial beef producers. Until recently these markets were considered the bastion of larger beef breeds but they are emerging as a rapidly growing new market for smaller breeds.

Response time

Do it now. If someone rings and ask for information do it within 24 hours. If they wanted a brochure in three weeks' time they would have asked for the brochure in three weeks. If you can't do it straight away, send them an email and acknowledge their enquiry or phone them and let them know when it will be sent. Then remember to follow up.

Internet

- It's there, it's huge and it's available so make the most of it. You don't have a choice.
- Get your own website. It does not have to be expensive to set up but you have to be part of it. Not one of our export markets would have been established without the web. Make it mobile-friendly.
- If you do develop a website then you must update it regularly. Consumer habits show that websites that update regularly will get repeat hits. Those that don't, won't.
- Get a presence on social media. Participate in groups that are relevant to the industry and your breed. I have made a number of export semen sales from advertising on these sites.

- Produce regular newsletters that can be sent to your clients.
- Most of our clients are internet savvy and spend a lot of time on it gathering information. They like it because it is immediate, so keep your site informative and change it regularly.

Marketing tips

- Don't miss a chance.
- Marketing is time-consuming but it does not have to be expensive. You just need to find ways to get your name/brand out there.
- Repeat sales are the best. According to the US Consumer Affairs department, it costs five times more to get a new customer than it does to retain one. Businesses that market to other businesses spend 4% of their revenue on marketing. Businesses that market to the general public spend 10%.
- Word-of-mouth referrals are invaluable. They are doing the work for you, as do people who show your stud stock. One satisfied customer is likely to tell five other people, while someone who is dissatisfied is likely to tell 20.

16

Useful equipment in small farms

These are some inventions and pieces of equipment that I either use on my farm or would like to use. I have no financial or sponsored interest in any of these products. I have not discussed the normal things like tractors or slashers as many farmers are very knowledgeable when it comes to their toys. My husband used to have a T-shirt that read, 'He who dies with the most toys wins'! The only piece of advice with regard to a tractor is to make sure it is over 70 hp if you want to do any serious farming. We started with hobby-farm tractors but they were hopeless: not strong enough to lift a round bale, drive a post-hole digger, move the grain bin, dig a hole to bury an animal or slash paddocks without overheating.

Fox lights

These are an Australian invention designed to keep foxes away from baby lambs but they are also very effective at keeping wild dogs away from small calves. I use one outside our chook yard to keep away the foxes, and a number are placed strategically in our paddocks. Since their installation we have had not a single dog attack. Fox lights give the impression of someone walking around with a flashlight. A computer chip on a powerboard with nine LED bulbs creates a varying flash of blue and white with different time sequences. It looks like there is a disco in your paddocks at night!

Gate magnets

The magnetic gate minder provides safety with the added efficiency of saving time when working with stock in yards. It is designed to hold a swinging gate fully open. This overcomes the possibility of stock getting in behind the gate, wasting time and effort. The magnet is strong enough to hold most swinging gates back against the rails but it retains the ability to release the gate quickly, allowing you to get the gate closed in a hurry. We use a couple in our yards and they are very effective.

Batch latch

The device can be adapted for many different uses. It is a portable solar-powered automatic gate-release timer, a New Zealand invention that was developed for the dairy industry to automatically allow cows in to feed. It's great for strip grazing or letting cows out of a yard after their feed if you have to go to work. I have adapted mine to automatically close the chook gate at night so I never have to remember to lock the chooks up at night before Mr Fox gets to them.

Cultivator Pro

The GreenPro Cultivator is a five-in-one farming implement which allows you to disc, plough, seed, cover and roll all in one pass. It can be towed behind an ATV.

Batch latch.

Cultivator Pro.

Doing all this in one pass means not driving over the seedbed laid by your discs, therefore there is less damage and soil compaction. It avoids having to buy five separate implements and is perfect for smaller acreage. It also saves time by eliminating the need for three or four implement changes and driving over the same piece of land several times.

Quick hitch

Ideal for people like me, who often don't have someone to help them. The quick hitch changes implements on your three-point linkage in literally seconds, no matter what size tractor you have.

Stock crates

There are a couple of companies that make excellent stock crates that can be towed behind your four-wheel-drive vehicle. This saves on the expense of having to buy a truck and get a truck licence. I have had one for 10 years and it has been amazing. We can cart our show team without the expense of paying a carrier, it has a sliding rear gate so we can load and unload from a loading ramp, it has a dropdown tail gate so we lead our cattle onto the back if there is no ramp or use some portable panels to make a race if need be, we unload directly into a paddock if clients don't have a ramp, drive the ATV onto it ... the uses are endless.

Mobile back rub frames

These are a great product to hold medicated back rub in the treatment of buffalo fly. They can be moved around easily from one paddock to the next. They have a towing attachment and a unique adjustable sliding hook.

Clipex star pickets

A wonderful invention that makes the easiest and fastest fencing system. It is a star picket with a clip mechanism that you clip the wires in. No more tying of wires! The benefit is that if you ever want take the fence down, for example to move something large through the paddock, you just unclip the wires then put them back up again when you're done.

Apps

There are some great apps to help you in your cattle production. They range from everything on soils, weather, pasture and types of production to cuts of beef. The list is endless and will continue to grow.

Software

For over 15 years I have used a program called Stockbook. It is integral to my business and has been the most incredible asset. It contains all the pedigrees of all animals and records all their treatments, weights, paddock movements, progeny, sales made, AI dates – just to name a few. I use it nearly every day and don't think I could run my business without it. I could not recommend it more highly. There are other programs on the market but this one works for me. I can integrate it with my scales which is particularly important so I can monitor weights for my carcass competitions. Also, when I sell cattle I can provide my clients with full health records detailing every vaccination and treatment, how much they cost, how much was used, the batch number, the withholding period, pregnancy status etc.

Q. I was wondering if you use a software program to manage your herd. Mine is proving to be a bit limited. I am looking for a system that will interface with the scales, and will ultimately allow data to be inputted from electronic readers in the yards.

A. There are a number of good software programs that allow you to record and keep track of your animals. The data can include pedigrees, date of birth, health treatments, weights, progeny. Purchase one as soon as you start then it will grow and expand with you, as does your herd. We use Stockbook that allows us to record all joining, health records, weights etc. It also interfaces with our scales and NLIS scanning.

17

Government regulations and laws

This section has been included as there are legal requirements, which must be observed if you don't want to incur hefty penalties, for owning, transporting and selling livestock. Each state has different requirements so check with your state government, but there are also mandatory national regulations.

Legal requirements

- Registration of your property with the Department of Primary Industries and Fisheries (DPI&F) will be necessary.
- Transportation or moving of livestock will require permits, waybills, National Vendor Declarations (NVDs), health permits and/or tick treatments.
- Animal identification regulations such as NLIS and branding will vary with each state.

Property identification code

Each property that wishes to have livestock must apply for a property identification code (PIC), issued by the state department of agriculture. If your land has previously had livestock, that number will be transferred to you upon purchase. If there has never been livestock on the property then you will need to apply for a property number. This will enable you to transfer cattle onto and off your property. You must maintain the accuracy of the contact details for the PIC allocated to you.

You must notify the DPI&F of the sale or transfer of your PIC. A PIC is also required to purchase NLIS devices.

NLIS

The National Livestock Identification System is Australia's system of identifying and tracing all cattle movements from property of birth to eventual sale, slaughter or export. Permanent NLIS-approved identification devices (i.e. ear tags or rumen boluses) must be used; these stay with the animal until death. **All cattle must be identified with an NLIS device before leaving a property.**

Data on movements of individual cattle between locations with different PICs is reported to and recorded in the NLIS database. It is your responsibility to report transfers within 48 hours of the completed movement, to keep the NLIS database current. You will need to provide the numbers in each individual animal's NLIS tag. To provide documentary information on animal movements, a waybill (NVD/waybill) is a legal requirement and must be held at origin and destination.

You will need to create an NLIS account: www.nlis.mla.gov.au.

Livestock Production Assurance program

This guarantees that the meat from your farm has been produced safely. Before you can obtain NVD forms you must register with the Livestock Production Assurance (LPA) program and commit to being an accredited producer.

When you tick the relevant box on your NVD form, you are guaranteeing that your on-farm practices meet LPA requirements. Here is an excerpt from my LPA.

> *[A] LPA COMMITMENT CONFIRMATION*
>
> *Attention:* **Margo Hayes,**
>
> *The LPA Database has recently been updated by you and shows that as the authorised representative of Property Identification Code (PIC)* **QILL0052,** *located at* **221 Mulgowie Rd, THORNTON QLD 4341,** *that you have confirmed your commitment to ensuring that the requirements of being an LPA Accredited Producer are understood and met at all times.*
>
> *A copy of the Declaration as agreed to by you is highlighted below for your records:*
>
> - *I understand that eligibility to use LPA National Vendor Declarations (LPA NVDs) is restricted to PICs that are accredited under the LPA Program;*
> - *I have read and understood the requirements of the LPA Program as described in the LPA Rules and Standards (as amended from time to time);*
> - *I will seek to ensure all persons with access to LPA NVDs for this PIC will comply with the requirements of the LPA Rules and Standards at all times;*

- *I will ensure that all records required by LPA Administration and the LPA Rules and Standards are maintained, including auditable evidence to demonstrate compliance with the five (5) elements of the LPA Standards: Property Risk Assessment, Animal Treatment records, Agricultural Chemical Usage and Stockfeeds, Preparation of Livestock and Livestock Movements and Transactions;*
- *I will provide access to LPA Administration to conduct random audits as required under the LPA program (irrespective of whether the business is a hobby farm, small/large operation, or is currently destocked);*
- *I will take any corrective and preventive action as required under the LPA Rules and Standards;*
- *I will cease using NVDs displaying the LPA logo if accreditation is withdrawn;*
- *All information provided to LPA applicable to this PIC is correct to the best of my knowledge;*
- *I will inform LPA Administration of all changes applicable to the LPA Accreditation of PIC QILL0052.*

Combined NVD/waybill

Combined NVD/waybills are available for cattle, bobby calves, sheep and goats from Meat & Livestock Australia. Every animal that is sold within Australia must be accompanied by an NVD.

NVDs were developed to assist producers to document the history of chemical use and treatment of animals offered for sale. The details assist processors and buyers who are seeking information on the history of sale stock. People completing NVDs are legally obliged to ensure that any information on their stock is completely accurate. Penalties apply if false or misleading information is given on the declaration.

The combined NVD/waybill has legal recognition in that it meets the legal requirements for a waybill. The combined NVD/waybills are produced in books of triplicate forms (one copy goes to the purchaser, one copy remains with the carrier and one copy is retained by the vendor for auditing purposes) and are available from Meat & Livestock Australia under the LPA program.

Also available is the e-DEC, which is a cheaper electronic alternative than traditional NVDs. The e-DEC allows producers to email NVDs directly to their buyers and print out NVDs from their computer.

Waybills (Qld)

Waybills are very rarely used any more, and the government has ceased printing waybill books. Waybills can be completed online and can only be obtained over the internet. Waybills describe the livestock being moved and are used when animals are only being transported from one place to another, i.e. no sale has occurred. Penalties can be imposed for incorrect use of waybills.

Transported Stock Statement (NSW)

TSSs must be used if an NVD is not. They are not required for transporting to a veterinary surgeon or to an event such as a show, field day etc. Copies must be kept for two years. Fines for non-compliance can be in excess of $5500.

Cattle Alternative Movement Waybill (SA)

While NVDs are preferred, a 'stand alone' movement waybill can be used as an alternative document for the purpose of meeting NLIS requirements in South Australia. NVDs are preferred for change of ownership of livestock.

Travel permits

Travel or stock permits assist in disease control and animal traceback. Most livestock movements require only an NVD/waybill but the following movements also require a travel permit. Travel permits are available from DPI&F offices or your local biosecurity officer:

- livestock travelling from cattle tick infected areas to cattle tick protected or free areas;
- where the property of origin is under quarantine for disease control or chemical residues;
- if stock are diseased or suspected of being diseased (a 'suspect' permit is required, e.g. for cancerous eye, lumpy jaw, enzootic bovine leucosis (EBL) etc.);
- livestock travelling to showgrounds;
- livestock travelling to interstate destinations;
- livestock travelling to export quarantine facilities.

In Queensland, there are severe penalties of up to $75 000 when stock are moved illegally into or out of cattle tick declared areas.

Transporting livestock

Under the *Australian Animal Welfare Standards and Guidelines: Land Transport of Livestock*,[9] any person in charge of an animal has a duty of care to maintain the animal's welfare before, during and after transport. This includes the person supplying the livestock for transport, the transporter and the person receiving the livestock at the final destination. The responsibility is shared by all involved in the process and all are accountable for their responsibilities.

Different states also have specific codes and guidelines. Victoria has the Code of Accepted Farming Practice for the Welfare of Cattle (www.depi.vic.gov.au/animalwelfare).

Queensland has introduced new laws for transporting livestock.[10] The compulsory code of practice commenced on 31 January 2014. The following key features are a synopsis of the rules and are in no way complete.

Livestock must be fit for travel

Livestock are not fit if they:

- are unable to walk bearing weight on all legs;
- are heavily pregnant (within four weeks of calving) and time off water is to be greater than four hours;
- gave birth within 72 hours prior to loading;
- are visibly dehydrated;
- are severely emaciated;
- are severely distressed or injured;
- are suffering a condition that is likely to increase their pain or distress during the transport process;
- are blind in both eyes.

Notifying time of arrival

Before transporting cattle, the person receiving the cattle must be notified of the estimated time of arrival.

Suitability of the transport vehicle

A suitable transport vehicle is one that:

- has adequate airflow appropriate to the type of cattle being moved;
- has suitable flooring that prevents slipping or falling;
- is free from objects or protrusions that could cause injury;
- is high enough to minimise injury;
- allows you to segregate unfamiliar groups and aggressive cattle;
- allows the correct load density.

Alignment of loading ramps

Both the transporter and the person responsible for loading the cattle must ensure that the vehicle and the ramp are properly aligned and close together to reduce the risk of injury to the livestock. NB: Even a small gap can cause a calf to break a leg.

Handling of livestock

This includes:

- no kicking, punching or striking of animals;
- no lifting or dragging by one leg;
- no electrical prodders on cattle less than three months of age;

Table 17.1: Regulations for transporting livestock

Class of cattle	Maximum journey time (hours)	Maximum time off water (hours)	Minimum spell duration (hours)
24–37 weeks pregnant	24	24	12
>37 weeks pregnant	4	4	24
>30 days of age but <6 months	24	24	12
Lactating cattle with dependent young	24	24	12
Any other cattle	48	48	36

- prodders cannot be applied to the face, udders, anus or genitals;
- restrict and muzzle biting dogs;
- don't use any dogs on bobby calves;
- only guardian dogs can travel in the same enclosure as livestock.

Calculating length of spell and time off water

Spell time starts from when the livestock are unloaded from the transport. Time off water starts from the time the livestock had reasonable access to water before starting the journey (Table 17.1).

When spelling cattle they should be unloaded, and given access to water and space to lie down. Feeding is not recommended for short stops of less than 12 hours. Detailed information on feeding cattle for each 24 hours they are at a facility are provided in the guidelines.

The laws aim to reduce the stress and potential injury risk to animals while being transported. Hopefully all producers will adhere to them, as there could be serious consequences for those who don't.

Brands and earmarks

The rules for each state are given in Table 17.2.

Table 17.2: State rules for brands and earmarks

Queensland	All cattle of 100 kg or more live weight must bear a registered brand before being sold. The brand must be registered
NSW	Branding is not compulsory but, if used, the brand must be registered
Victoria	Branding is not encouraged and no formal brand registration system is maintained
SA	Branding is not compulsory but, if used, the brand must be registered
WA	Branding or earmarking is compulsory for all cattle and the brand and earmark must be registered
Tasmania	Branding is covered by legislation and therefore compulsory but it is not enforced. If used, the brand must be registered
NT	All cattle over eight months of age must be branded before leaving the property and the brand must be registered

Animal health and welfare

New Australian Animal Welfare Standards and Guidelines are set to come into effect in 2015. They cover all aspects of care from responsibility to feed and water, facilities, handling, castration and dehorning to breeding management. They also include dairies, feedlots and humane killing methods.

Duty of care

People who keep animals have a 'duty of care' regarding their welfare. Provision of health management, nutrition and housing is required by state legislation. Acts of cruelty are prohibited. Model Codes of Practice for the care of most species are available from the CSIRO. The codes set out the minimum acceptable standards for the keeping of livestock.

Nutrition

It is illegal to feed cattle or other ruminants with feed containing restricted animal material which is defined as blood meal, meat meal, meat and bone meal or other meal made by rendering material of animal origin, excluding gelatin, milk of Australian origin, milk products made in Australia from Australian milk, and tallow. It also bans the feeding of household scraps such as salami, cheese and ham, due to the risk of these products carrying exotic disease viruses. This requirement was introduced to protect Australia against exotic diseases such as mad cow disease (BSE) and foot-and-mouth disease (FMD). It is strictly enforced.

Livestock on public roads

In all states of Australia, except Queensland and the Northern Territory, the common law rule known as Searle V Wallbank, under which owners of stock are under no legal obligation to maintain fencing or to exercise care to prevent animals straying onto a road, has been abolished by statute. In all other states of Australia landholders with a property adjoining a road have a duty of care to the users of that road. Research by the Western Australian Pastoralists and Graziers Association in 2104[11] on road accidents involving livestock found 28 accidents were caused by cattle, 280 were caused by horses, 383 accidents were caused by kangaroos and 279 accidents involved unknown livestock.

Chemical residues

The National Organochlorine Residue Management (NORM) program is supported by all state Departments of Primary Industries and Fisheries (or equivalent). Animal products are monitored closely to detect any residues from the use of chemicals. Residues may cause loss of markets for livestock products. This is of particular significance to the beef industry.

All DPI&F-registered farms are allocated an organochlorine residue status which is listed on the national Extended Residue Program (ERP) database, which is maintained and managed by Meat & Livestock Australia. Farms that hold a higher risk status are subject to significant testing at abattoirs. This may be at the vendor's expense.

New property owners, when obtaining a PIC and registering with the LPA program, will be able to ascertain their status prior to selling cattle. The LPA program assists producers to ensure that unacceptable chemical residues and contaminants do not occur in livestock presented for slaughter.

Observing the following rules will prevent residue problems:

- use registered chemicals ONLY according to label instructions;
- read and follow the instructions on the container before using or applying any agricultural or veterinary chemical;
- observe withholding periods;
- observe export slaughter intervals.

Careless use of chemicals can result in contamination of livestock and their products (milk, meat, eggs and fibre). Waste vegetables and fruit may contain chemical residues. Livestock owners are responsible for making sure that their animals do not have access to contaminated feed.

Withholding periods and export slaughter intervals

A product's withholding period (WHP) is the legal period between treatments and slaughter for domestic markets. It is the length of time required to ensure that any chemical residue has fallen below the maximum residue limit (MRL) at product harvest. MRLs are the maximum chemical concentrations that are permitted in foodstuffs. WHPs are legally binding and are printed on chemical/drug labels.

WHPs and export slaughter intervals (ESIs) can differ. ESI is the interval between treatment and slaughter for export product. It refers to voluntary export guidelines designed to ensure chemical residues fall below minimum overseas requirement levels, as these vary from country to country. Livestock producers must ensure that any animals made available for slaughter comply with WHPs and ESIs.

Tip: Tables for ESIs and WHPs for nearly all chemicals are found on the reverse side of the NVD form.

Glossary

Acre a unit for measuring land. 1 acre = 0.405 hectares
Anoestrus the non-breeding period where there is no evidence of oestrus
Bobby calf a calf that is sold straight off its mother soon after birth
Bos indicus the species of cattle which are referred to as tropical or humped breeds
Bos taurus the species from which temperate, British and European breeds of cattle have developed
Bull an entire male bovine of any age
Bulling a cow which is on heat
Bullock a castrated animal about two years of age or more
Calf a young animal of either sex from birth to weaning age
Carcass weight once slaughtered, this is the weight of the carcass less all the parts that cannot be sold as meat to the butcher (hide, head, intestines, hooves)
Composite breed a new breed developed by crossing two or more breeds then stabilising that cross. Many breeds that are called purebreds are actually composites, e.g. Mandalong specials
Cow a female animal of breeding age
Dam a parent female
Dark cutting the dark red or black colour of carcass meat which usually results from stress before slaughter
Dressing percentage Carcass weight as a percentage of the live weight at slaughter:

$$\frac{\textit{Carcass weight}}{\textit{Final weight}} \times 100$$

Dry cow a cow that is not giving milk
Dystocia calving difficulty
EBV estimate breeding value
Electrolytes a substance containing free ions that behaves as an electrically conductive medium
EMA eye muscle area
Embryo transfer technique of removing an embryo from one female and implanting it in another

Embryo an organism in the early stages of development in the uterus
ESI export slaughter interval
Fitter someone who shows and parades cattle professionally
Freemartin a generally sterile heifer calf twinborn with a bull
Genetics the science of inheritance and the way genes are passed from one generation to another
Genotype the genetic makeup of an animal
Hectare a unit for measuring land. 1 hectare = 2.47 acres
Heifer a young female before she has had her first calf
HGP hormonal growth promotants. Implanted to produce faster growth rates in cattle, particularly used in feedlots
Hybrid vigour heterosis (difference in performance of vigour of crossbred progeny)
Joining the term used when a male is joined to a female
Led and tractable animals that are halter-broken and can be led
Live weight the weight of an animal
MLA Meat & Livestock Australia
MSA Meat Standards Australia
NLIS National Livestock Identification System
NVD National Vendor Declaration
Phenotype the physical makeup of an animal, which is a result of interaction between the genotype and environment
Polled a genetically hornless animal
Retail yield the percentage of saleable meat (once the bone and excess fat have been removed)
Sanga *Bos taurus* breeds that have developed tropical adaptation traits
Scours the term for diarrhoea
Sire a parent male
Springer a cow that is about to give birth
Stag a bull that has been castrated after or upon reaching sexual maturity
Steer a castrated male
Supplements additional feeds or products usually given to rectify a deficiency
Trait character or characteristic
Vealer a calf of either sex still drinking milk
Weaner a calf separated from its mother
WHP withholding period
Yearlings cattle that are approximately 12 months of age

Cattle organisations

Ausline Cattle Association
PO Box 6144, Buranda Qld 4102
Ph: 1300 882333
Email: info@auslinecattle.com.au
www.auslinecattle.com.au

Australian Galloway Association
PO Box 42, Westbury Tas. 7303
Ph/fax: (03) 6393 2866
Email: office@galloway.asn.au
www.galloway.asn.au

Australian Highland Cattle Society Inc.
C/- ABRI, University of New England, Armidale NSW 2351
Ph: (02) 6773 3144
www.highlandcattle.org.au

Australian Lowline Cattle Association
C/- ABRI, University of New England, Armidale NSW 2351
Ph: (02) 6773 3295
Email: info@lowlinecattleassoc.com.au
www.lowlinecattleassoc.com.au

Australian Miniature Hereford Breeders Network
email: amhnetwork@gmail.com
Ph: Julie 0418 484 888
www.amhbn.com

Australian Miniature Hereford Cattle Association
PO Box 415, Waroona WA 6215
Secretary: secretary@miniatureherefords.org.au
www.miniatureherefords.org.au

Boran Cattle
Ph: 0448 523 618
Email: president@austsmallboran.com.au
www.austsmallboran.com.au

Bramalow Breeders' Group
Email: coleglenstud@hotmail.com or leonalabrie@gmail.com

Dexter Cattle Australia Inc.
University of New England, Armidale NSW 2351
Ph: (02) 6773 3471
Email: dexter@abri.une.edu.au
www.dextercattle.org.au

Nadudana Cattle Queensland
President: (07) 5435 0289
Vice President: (07) 4947 2596
www.nadudanacattleqld.com.au

Square Meaters Cattle Association of Australia Ltd
PO Box 189, Kiama NSW 2533
Ph: (02) 4232 3333
Email:squaremeaters@bigpond.com
www.squaremeaters.com.au

Further information

Recommended websites

Animal Welfare standards: www.animalwelfarestandards.net.au
Australian Model Code of Practice for the Welfare of Animals (Cattle): www.publish.csiro.au
Department of Primary Industries and Fisheries (Qld): www.daf.qld.gov.au
Department of Primary Industries (NT): www.nt.gov.au
Department of Primary Industries (NSW): www.dpi.nsw.gov.au
Department of Primary Industries and Resources (SA): www.pir.sa.gov.au
Department of Primary Industry, Water and Environment (Tas.): www.dpiw.tas.gov.au
Dr Sandi Jephcott (beef cattle veterinary and nutrition consultant): sandibeef@gmail.com
Dr Temple Grandin: www.grandin.com
Environment ACT
Farming Ahead: www.farmingahead.com.au agricultural portal
FarmOnline: www.farmonline.com.au
Farmsafe Australia: www.farmsafe.org.au
FutureBeef (an excellent resource centre for northern Australia): www.futurebeef.com.au
http://www.healthysoils.com.au
http://www.soilfoodweb.com.au
http://www.soilhealth.com
Meat & Livestock Australia: www.mla.com.au
Pinkeye information: www.pinkeye.com.au

Useful publications

Beef Central (a daily source of beef cattle news): www.beefcentral.com

Blues Country Magazine (monthly magazine): www.bluescountry.com.au
FarmStyle Australia (specialised small farm advice): www.farmstyle.com.au
Rural Press has state-wide weekly newspapers
Small Farms Magazine (a monthly farming publication full of excellent articles): www.smallfarms.net
Smart Farmer (a very useful bi-monthly magazine supplement to Queensland Country Life): www.queenslandcountrylife.com.au

Equipment

Allflex (management tags): www.allflex.com.au
Animal Health Direct (online supplier for all livestock supplies): www.specialistsales.com.au
Arrow (yards and equipment): www.arrowfarmquip.com.au
Biopryn System (pregnancy testing system): www.pregtest.com.au
Clipex (fencing supplies): www.clipex.com.au
Easywean (weaning): www.easywean.com.au
FarmBitz (quick hitches): www.farmbitz.com
Gallagher (cattle scales): www.gallagher.com.au
GrazeTech (batch latches): www.grazetech.com.au
Jarvis Fabrication (mobile back rubs): www.jarvisfabrication.com.au
KaraKar (cattle crates): www.karakar.com.au
Kline Irrigation (portable irrigation system): www.k-line.com.au
Leader (management tags): www.leaderproducts.com.au
Metalcorp (yards and equipment): www.metalcorpsteel.com.au
Mi-Feed (micronised feed solutions): www.mi-feed.com.au
Practical Systems (Stockbook software programs): www.practicalsystems.com.au
QuikCorp (GreenPro Cultivator Pro): www.quikcorp.com.au
Ridley Feeds (stud cattle feeds): www.agriproducts.com.au
Shoof (farmers' catalogue): www.shoof.com.au
The Cattle Shop (excellent supply of shopping needs): www.thecattleshop.com.au
The Farmers Mailbox (fantastic one-stop shop for all your needs): www.fmb.com.au
The Stocker (cattle crates): www.nationaltrailers.com.au
Think Livestock (vaccination supplies): www.thinklivestock.com
Thunderbird (cattle scales): www.thunderbird.net.au
Titan Trailers (cattle crates): www.usahorsefloatsaustralia.com.au
Tru-Test (cattle scales): www.trutest.com.au
University of Queensland (DNA testing) http://www.uq.edu.au/vetschool/cattle
Virbac (vaccines): www.virbac.com.au
Zoetis (vaccines and DNA testing): www.zoetis.com.au

Supplements and feeds

Anipro (molasses supplement): www.anipro.net

International Animal Health (excellent range of health products): www.iahp.com.au

Courses

All State Ag (AI courses): www.allstateag.com.au
RCS (grazing for profit schools): www.rcsaustralia.com.au
Vitulus (introductory cattle care courses): www.vitulus.com.au

Apps

Forage Budget: www.stocktakeplus.com.au
Measure Your Land: iTunes
MLA Market Information: latest livestock market information
SoilMapp

Recommended reading

Beef Cattle Nutrition: An Introduction to the Essentials. Publication Code1740369289, MLA website.

Bertram J (1995) *Bull Selection*. Queensland Department of Primary Industries, Brisbane.

Boothby D & Fahey G (1995) *Artificial Breeding of Cattle*. Agmedia, Melbourne.

Brightling A (1994) *Stock Diseases*. Inkata Press, Sydney.

Dowling R & McKenzie R (1993) *Poisonous Plants: A Field Guide*. Publication Q192035. Queensland Department of Primary Industries, Brisbane.

Entwistle K & Fordyce G (2003) *Evaluating and Reporting Bull Fertility*. AACV, Brisbane.

Hungerford TG (1990) *Diseases of Livestock*. 9th edn. McGraw Hill, New York [the veterinary bible].

Newham L (1994) *Beef Cattle: Breeding, Feeding and Showing*. Inkata Press, Sydney.

Powell E & Lapworth J (2002) *Cattle Yards: Design, Materials and Construction*. 2nd edn. Queensland Department of Primary Industries, Brisbane.

Wilson J (1991) *Farm Cattle Handling SCA Report No 35*. CSIRO.

Wilson J (1994) *Beef Cattle Judging*. Agricultural Societies Council of NSW.

Wilson J (2001) *Keeping a Cow*. NSW Agriculture, Tocal.

Suggested MLA tools and resources

Cost of production tool: http://www.mla.com.au/News-and-resources/Tools-and-calculators/Cost-of-production

Feed budget and rotation planner: http://www.mla.com.au/News-and-resources/Tools-and-calculators/Feed-budget-and-rotation-planner

Feed demand calculator: http://www.mla.com.au/News-and-resources/Tools-and-calculators/Feed-demand-calculator

Pasture health kit: Publication Code 1740366751.
Pasture improvement calculator: http://www.mla.com.au/News-and-resources/Tools-and-calculators/Pasture-improvement-calculator
Rainfall to pasture growth outlook tool: http://www.mla.com.au/News-and-resources/Tools-and-calculators/Rainfall-to-pasture-growth-outlook-tool

Endnotes

1 Brew MN, Carter J & Maddox MK (2009) *The Impact of Water Quality on Beef Cattle Health and Performance*. Publication AN#187. University of Florida. https://edis.ifas.ufl.edu/an187

2 http://www.business.qld.gov.au/industry/agriculture/animal-management/cattle/managing-tick-fever-in-cattle/natural-tick-fever-immunity

3 Drench resistance: a growing problem. *Queensland Country Life*. 9 October 2014, p. 68.

4 Arthur P (2014) Protein push. *Queensland Country Life*. 11 December 2014, p. 54.

5 *Pasture Growth: Build and Maintain Soil Nutrients to Improve Soil Fertility and Health in all Pasture Zones*. http://www.mla.com.au/mbfp/Pasture-growth/3-Build-and-maintain-soil-nutrients

6 MLA. *Livestock Production: toxic plants*. mla.com.au

7 Australian Organic Market Report 2014, p. 5. Australian Organic Ltd.

8 Australian Grainfed Minimum Standards Specifications: Aus-Meat.com.au

9 *Australian Animal Welfare Standards and Guidelines: Land Transport of Livestock*. Edition 1, Version 1.1, 21 September 2012.

10 Queensland *Animal Care and Protection Act 2001*, Animal Care and Protection Regulation 2012.

11 Garnett O (2014) Cars v. livestock: who's liable. *ABC Rural*. 11 June 2014.

Index

www.ingramcontent.com/pod-product-compliance
Lightning Source LLC
LaVergne TN
LVHW080043210926
848481LV00011B/3263
* 9 7 8 1 4 8 6 3 0 1 8 6 7 *